Stability Analysis of Regenerative Queueing Models

Evsey Morozov · Bart Steyaert

Stability Analysis of Regenerative Queueing Models

Mathematical Methods and Applications

Evsey Morozov
Institute of Applied Mathematical Research
Karelian Research Centre
Petrozavodsk, Russia

Bart Steyaert
Department of Telecommunications
and Information Processing
Ghent University
Ghent, Belgium

ISBN 978-3-030-82440-2 ISBN 978-3-030-82438-9 (eBook)
https://doi.org/10.1007/978-3-030-82438-9

This Springer imprint is published by the registered company Springer Nature Switzerland AG
The registered company address is: Gewerbestrasse 11, 6330 Cham, Switzerland

To beloved Irina, Anna, Taisia, Viktor and unforgettable Mikhail

Evsey Morozov

To Maayken, with all my heart

Bart Steyaert

Abbreviations, Notations and Conventions

Commonly Used Abbreviations

iid	independent and identically distributed;
PASTA	Poisson arrivals see time averages;
SLLN	Strong law of large numbers;
w.p.1	with probability 1;
FIFO	First-In-First-Out service discipline;
FCFS	First-Come-First-Served service discipline;

Commonly Used Notations and Conventions

$\mathsf{P}(A)$	probability of the event A;
$1(A)$	the indicator function of the event A;
$\mathsf{E}X$	the expected value of the random variable X;
$F(x)$	(cumulative) distribution function;
$\bar{F}(x) = 1 - F(x)$	tail distribution function;
$X =_{st} Y$, $X \leq_{st} (\geq_{st}) Y$	stochastic relations between the random variables X, Y;
$\Rightarrow$	convergence in probability/distribution;
$\to$	convergence w.p.1;
t_n	the arrival instant of customer n;
τ	a generic interarrival time, with arrival rate $\lambda = 1/\mathsf{E}\tau \in (0, \infty)$;
S	generic service time with service rate $\mu = 1/\mathsf{E}S \in (0, \infty)$;
$S(t)$	remaining service time at instant t;
$A(t)$	the number of arrivals during the time interval $[0, t]$;
$D(t)$	the number of departures during the interval $[0, t]$;
$V(t)$	the workload that has arrived during $[0, t]$;
$\nu(t)$	the number of customers in the system at time t;

$Q(t)$ the number of customers in the *queue* (the system without servers) at time t;
$W(t)$ the remaining work at time t in a single-server system;
$\mathrm{W}(t)$ the total remaining work at time t in a multiserver system;
$I(t)$ the idle time during $[0, t]$;
$B(t)$ the busy time during $[0, t]$;
P_0 the stationary idle probability;
P_B the stationary busy probability;
T a generic regeneration period in continuous time;
θ a generic regeneration period in discrete time;
$(x)^+$ $= \max(0, x)$;
$\sum_{\varnothing}$ $= 0$.

The above notations can be further extended or adapted, if appropriate, for the model or analysis that is considered. In particular, in the above notation an index i will be added to refer to server i (in a multiserver system) or to a class-i customer (in a multiclass system). Moreover, in order to denote a generic element of an iid sequence of random variables, we omit the serial index.

At this point it is also useful to notify the reader that the convergence 'w.p.1' is often assumed *implicitly* when we consider the convergence of random variables, and the limiting fraction of customers or time is sometimes treated as a *stationary probability*.

Theorems, remarks and problems are numbered within each chapter, for instance, Theorem 2.7 refers to the 7th theorem within Chap. 2

The list of papers that we mention at the end of each chapter is by no means exhaustive, but meant to be a representative sample of interesting papers in the related scientific area.

Contents

Chapter 1
Introduction

The stability analysis of stochastic models for telecommunication systems is one of the topics intensively studied nowadays. This analysis is, as a rule, a difficult problem requiring a refined mathematical technique, especially when one endeavors beyond the framework of Markovian models. Stability analysis allows to delimit the region of the system parameters where the stationary regime of the processes describing the dynamics of the associated queueing system exists. The method that is presented in this book is devoted to the stability analysis of queueing systems possessing the *classical regeneration property*. This means that a trajectory of the random processes describing the behavior of the system can be divided into iid *regeneration cycles* at the consecutive *regeneration instants* which constitute an *embedded renewal process*. The dynamics of a wide class of queueing models describing modern communication and computing systems can be expressed in terms of such regenerative processes [1–7].

An important feature of the presented approach is a separation of the predefined conditions on the *negative drift condition* and the *regeneration condition*. The former condition requires that the mean increment of the workload process (the remaining amount of work in the system at any given time instant) or queue size process outside a bounded set B (including a *regeneration state*) must be negative. This, in turn, implies that the process itself *does not go to infinity* and thus visits the set B infinitely often. As an alternative, the latter (regeneration) condition is used to show that, starting in the set B, the process starts a new regeneration cycle within a finite time interval with a positive probability. Then we use a key asymptotic result from renewal theory to establish the system's stability. More precisely, we use a characterization of the limiting behavior of the remaining regeneration time, or, in other words, the remaining renewal time of the renewal process of regenerations. This characterization claims that if the length of the remaining renewal time does not go to infinity in probability, then the mean length of the regeneration period (the mean distance between the adjacent regeneration instants) is finite [8]. Such a regenerative process is called *positive recurrent*, by analogy with a Markov process. From the

E. Morozov and B. Steyaert, *Stability Analysis of Regenerative Queueing Models*,
https://doi.org/10.1007/978-3-030-82438-9_1

point-of-view of practical applications, a queueing system described by a positive recurrent regenerative process can also be called *stable* or *stationary*. As follows from the *asymptotic theory* of regenerative processes, it turns out that positive recurrence (under a mild technical assumption) implies the existence of the stationary distribution of the process under consideration [1, 3]. This observation allows us to further reduce the stability analysis by verifying that, under predefined negative drift and regeneration conditions, the queueing process *does not go to infinity*. In many cases this verification turns out to be much easier than a direct proof of the finiteness of the mean regeneration period.

Although the stability conditions of the classical queueing systems that we present below (in particular, in Chaps. 2 and 4) are well known, their analysis allows us to demonstrate our method in great detail, and outlines an 'algorithm' on how to act in the stability analysis of other models. An important element of the analysis is that we also investigate several models under an *arbitrary initial state* in which case the first regeneration cycle has a different distribution than a generic cycle (starting at a regeneration point). In this case, positive recurrence also requires the finiteness w.p.1 of the first regeneration period. The main underlying idea of the analysis here is the following. Under a negative drift condition, we prove that the basic regenerative process hits a bounded set B *infinitely often* w.p.1, while, under the regeneration condition, we show that the mean number of such visits is *finite within each regeneration cycle*. This immediately implies that the number of regeneration cycles is not less than two, and hence the first period is finite w.p.1.

Besides a thorough study of the stability properties of various queueing models, this work also contains elements of the *performance analysis* of these queueing systems, under the condition that a stable regime has been reached. In many cases these performance measures are a byproduct of the stability analysis, in other cases, an explicit expression of a performance measure directly implies the corresponding stability conditions.

An important ingredient of this work is a large number of simple *problems*, which extend the presented material, and help the reader to develop some basic skills required for the regenerative stability analysis of queueing systems and related research areas.

Although this text primarily is oriented towards the mathematical aspects of the regenerative stability analysis method, the authors believe that the material is also of interest to the engineering community working in the telecommunication field, who may be faced with the problem of stability of queueing systems. As the experience of the authors shows, such problems, being quite common, attract a wide interest of researchers, but often turn out to be very challenging and difficult to be solved by conventional methods used by engineers for these purposes, say, in the framework of Markov models. On the other hand, the authors believe that this book can be useful also for 'applied mathematicians' (or more precisely 'applied probabilistic' researchers), because this book can be considered as a bridge between rather advanced mathematical books (as [1, 9–15]) and 'engineering' books (e.g. [16–20], just to mention a few) that sometimes suffer from a lack of mathematical profoundness in the analysis. Therefore, we have attempted to streamline our mathematical techniques with this audience in mind.

Summarizing, our main approach is, generally speaking, as follows: in order to prove the necessary stability conditions for a given queueing system that is first described in detail, we deduce the main balance relations between the amount of arrived and departed work in a given time interval, and then apply convergence in distribution and w.p.1, including the SLLN, in addition to asymptotic results for renewal and regenerative processes. As a rule, this approach not only leads to the necessary stability conditions, but also to relations between stationary performance measures, which in some cases lead to explicit expressions for these quantities. Moreover, we sometimes apply a *coupling method* allowing to consider different random variables defined on the same probability space, and as a result, compare the corresponding performance measures; in some cases this provides bounds for these quantities. In addition, when deducing the sufficient stability conditions, we often assume by contradiction that, under predefined conditions, the system is not positive recurrent and then consider the so-called *saturation regime* when the system is analyzed provided that the basic queueing process *goes to infinity*. In many cases such a saturated system turns out to be easier to analyse. This then allows us to prove that the 'saturated regime' assumption contradicts the predefined assumptions and therefore yields sufficient stability conditions.

The primary purpose of the authors is to present, in a unified form, their own experience in the research area that focuses on the stability analysis of regenerative queueing systems, although not claiming to describe the entire field of the regenerative approach, or other known effective methods applied to the stability analysis of queueing systems. We also would like to notify the reader that this book does not treat issues related to the rate of convergence to stationarity, which is an important but separate problem closely related to stability analysis.

For the interested reader, we recommend the survey paper [6] which, both in an intuitive and in a more formal way, gives a remarkable introduction to the theory of regenerative processes with many simple and instructive examples. For a deeper study of this theory we can recommend the fundamental books [1, 3, 7].

Finally, we would like to express our gratitude to many colleagues who influenced our research or collaborated with us. EM would first of all like to pay tribute to Professors Alexander Borovkov, Boris Gnedenko and Igor Kovalenko, whose research decisively inspired him throughout his scientific work. Also, EM would like to primarily mention Søren Asmussen, Vladimir Kalashnikov, Vladimir Rykov, and Richard Serfozo, whose research had a significant influence, among other things, on the development of the regenerative stability analysis method presented in this book. Moreover, EM is grateful to Jesus Artalejo and Konstantin Avrachenkov, for attracting his interest towards the stability of retrial queueing systems, which are given a significant place in this book. BS would like to foremost mention Herwig Bruneel, in addition to Koenraad Laevens, Sabine Wittevrongel, Joris Walraevens, Dieter Fiems, and Dieter Claeys, and thank them for the past fruitful long-term cooperation in the field of performance studies of queueing systems.

Moreover we are grateful to Konstantin Avrachenkov, Dieter Fiems, Ingemar Kaj, Masakiyo Miyazawa and Ilkka Norros for some useful comments on the draft version of this book. Our special thanks goes to Richard Serfozo and the anonymous reviewers for a number of useful suggestions and comments that resulted in helpful improvements of the draft version of the book. While we are grateful to colleagues for the improvements that were suggested, only the authors are responsible for the remaining shortcomings of the book.

The work of EM is supported in part by Moscow Center for Fundamental and Applied Mathematics, Moscow State University.

1.1 Structure of the Book

Let us now briefly outline the structure of the book.

In the remaining part of this chapter we present some basic concepts and properties related to renewal theory and the theory of regenerative and cumulative stochastic processes, which are widely used in the chapters that follow. Moreover, the coupling method, which plays an important role in some parts of the analysis presented in the following chapters, is briefly discussed.

In the 2nd Chapter, we demonstrate the main steps of the regenerative stability analysis by first considering the classical $GI/G/1$ and $GI/G/m$ queueing systems. The stability conditions of these system are indeed well known (see, for instance, [1, 21]), and the proofs that are developed in this chapter illustrate in detail the main generic steps of the regenerative stability analysis method. Moreover, these steps reappear under various forms when we consider other, more complicated queueing models. Although a single-server queueing system is a particular case of the corresponding multiserver system, we believe that it is instructive to increase the complexity on a step-by-step basis in the development of the stability analysis method, and therefore to consider the cases $m = 1$ and $m > 1$ separately. For the same reason, we make a distinction between the zero initial state multiserver system and the one with an arbitrary initial state. Also, some results with respect to instability of these systems are included. In addition, at the end of this chapter, using the regenerative approach, we deduce known explicit expressions for some of the basic stationary performance measures as well.

In Chap. 3, we establish the *tightness* property of some associated processes, which play an important role in the stability analysis developed in Chap. 2, and the following chapters as well. In particular, we focus on the tightness of the remaining service time process, and establish a delicate connection of this process with the well-studied remaining renewal time process. It is worth mentioning that the tightness of the remaining service time process even holds in the case of *unstable systems*. In addition, we establish the tightness of the *range* (being the difference between max and min) of the components of the *Kiefer–Wolfowitz* workload vector process in a more general setting than was considered in the keystone paper [21].

Chapter 4 contains the stability analysis for some *generalized many-server systems*, including multiserver and multiclass systems, an infinite-server and loss system, and a related finite-buffer system. A multiserver system with regenerative input is also considered. In this chapter we also present a number of stationary performance characteristics of these systems in an explicit form, mainly well-known results, but also some lesser-known ones.

In Chap. 5, we study the stability of *state-dependent systems*. In particular, in these systems there exists a mechanism allowing the control of some key processes, depending on the current system state, or in other words, these systems have a *feedback* between their system state and the governing 'external' processes, such as the input process or/and service time process, among others. The analysis of state-dependent systems nicely illustrates an important feature of the regenerative stability analysis: in order to establish positive recurrence, it is sufficient to show the negative drift of a basic queueing process when the system approaches a saturated regime.

In Chap. 6, we develop the regenerative stability analysis for some complicated systems with interacting servers. Such a queueing system configuration has a lot in common with the notion of systems with *flexible servers*, where some service capacity may be transferred from one pool of servers to another to accommodate varying demands. More specifically, we focus on the so-called N*-models* in which customers may trickle down from one station to another, but are not allowed to move up in the opposite direction.

Chapter 7 is devoted to the stability analysis of a *multiclass retrial* system with *constant retrial rates*, in which the retrial rate of a customer that is blocked in a virtual orbit only depends on the class of the customer, but not on the size of the orbit. In addition to sufficient and necessary stability conditions, we also present an alternative approach to the stability problem by means of a *loss dominating system*.

In Chap. 8 we consider *coupled orbit queues*, in which the customer retrial rate depends both on the customer class and on the binary states of all other orbits: either *busy or idle*. Therefore, the retrial rate of an orbit depends in general on a vector state of all orbits with binary components, called the *configuration*. It is worth mentioning that, in Chaps. 7 and 8, a key element of the proofs is based on a coupling argument, relating the process of retrial attempts from an orbit to an independent Poisson process. Then after using the PASTA property, we deduce the necessary stability conditions, as well as explicit expressions for some stationary probabilities.

In Chap. 9, we focus on the stability of another important class of multiclass retrial systems, namely *classical* retrial systems, where the retrial rate depends on the customer class, and is also proportional to the orbit size. The stability analysis exploits the property that the service discipline in this system is *asymptotically non-idling*, and the retrial system therefore approaches a classical buffered system as the workload process increases.

Finally, in Chap. 10, without going into too much detail since similar aspects of the analysis have been treated in the preceding chapters, we outline the regenerative stability analysis of an optical buffer, a discrete-time queueing system with renewal-type interruptions, and the proof of the necessary stability conditions of regenerative queueing networks.

1.2 Regenerative Processes

We will not present a comprehensive introduction to the theory of regenerative processes, which can be found for instance in [1, 3, 22], among others. Instead, we confine this section to a few basic definitions and results which are used in the analysis developed in the subsequent chapters in this book.

Intuitively, a stochastic process is *(classically) regenerative* if its path can be split into iid random elements called *regeneration cycles*, and the points separating these cycles are called *regeneration instants*. The simplest example of a classically regenerative process is an irreducible Markov chain with countable state space. In this case, as regeneration instants we can take the times when the chain returns to a fixed state, and the trajectory of the chain between two adjacent regeneration instants is a regeneration cycle. Another example, which is especially important in the context of the analysis that follows, is the $GI/G/1$ queueing system with iid interarrival times and iid service times. In this system, a regeneration of the queue size and the workload process occurs when a newly arriving customer observes a completely idle system. At each such an arrival instant, these processes start anew and independently of their history up to this instant. Moreover, they have the same probabilistic behaviour after each regeneration point. It then easily follows that the obtained regeneration cycles are indeed iid.

To be more formal, consider a continuous-time stochastic process $\{X(t),\ t \geq 0\}$ with a metric state space (for instance, the m-dimensional Euclidean space $\mathbb{R}^m$). With respect to the stochastic processes that we consider below, it is standard practice to assume that the process paths are *right-continuous with left-hand limits* (also called the *càdlàg* function), see [1, 3, 6].

On the same probability space as the process $\{X(t)\}$, suppose there exist random times $0 = T_0 < T_1 < T_2 < \cdots$ such that $T_n \to \infty$ w.p.1. We consider the process $\{X(t)\}$ in terms of its cycles on the time intervals $[T_n,\ T_{n+1})$. Specifically, the nth cycle of $\{X(t)\}$ is denoted by its nth *cycle process*

$$G_n := \{X(t) : T_n \leq t < T_{n+1}\}, \quad n \geq 0\,,$$

with cycle length $T_{n+1} - T_n$. The random times $\{T_n\}$ will be called regeneration instants, and the process $\{X(t)\}$ can be completely described by its cycle sequence $\{T_{n+1} - T_n,\ G_n\}$, called regeneration cycles, provided that these quantities meet the following definition:

Definition 1.1 The process $\{X(t)\}$ is regenerative over the time instants $\{T_n\}$ if its cycles, $\{T_{n+1} - T_n,\ G_n,\ n \geq 1\}$, are iid and independent of the initial cycle $\{T_1,\ G_0\}$.

When applying regenerative analysis, it is useful to understand that the process $\{X(t)\}$ is (classically) regenerative if and only if

(i) $\{X(T_n + t),\ t \geq 0\}$ is independent of T_n and the prehistory, $\{X(t),\ t < T_n\}$, $n \geq 1$;

(ii) the distribution of $\{X(T_n + t),\ t \geq 0\}$ is independent of $n \geq 1$.

The process $\{X(t)\}$ is called *delayed* if the initial cycle has a different distribution than the other cycles, and is called *zero-delayed* otherwise. In the zero-delayed case, T_1 is distributed as any time period $T_{n+1} - T_n$, and we denote by T a generic regeneration period. Define the *remaining regeneration time* at instant t as

$$T(t) = \min_{n\geq 1} (T_n - t : T_n > t), \quad t \geq 0\,. \tag{1.2.1}$$

Note that, in the zero-delayed case, $T(0) = T_1 =_{st} T$. One of the key elements of the approach that we present below is the following asymptotic property (Chap. XI, [8]): if $\mathsf{E}T = \infty$, then

$$T(t) \Rightarrow \infty\,, \quad t \to \infty\,, \tag{1.2.2}$$

or equivalently, for each $x \geq 0$,

$$\lim_{t\to\infty} \mathsf{P}(T(t) > x) = 1\,. \tag{1.2.3}$$

Moreover, if $T_1 < \infty$ w.p.1, then $T(t) \Rightarrow \infty$ implies $\mathsf{E}T = \infty$.

The following notion is a key one in the stability analysis which is developed in this book.

Definition 1.2 The process $\{X(t)\}$ is called *positive recurrent* if

$$\mathsf{E}T < \infty \quad \text{and} \quad T_1 < \infty \text{ w.p.1.} \tag{1.2.4}$$

We would like draw the reader's attention to the fact that the assumption $T_1 < \infty$ allows to ignore the $1st$ regeneration cycle in the asymptotic analysis of the regenerative process.

Remark 1.1 According to Definition 1.2, in a *non-positive recurrent* system at least one of the two assumptions, $\mathsf{E}T < \infty$ and $T_1 < \infty$ w.p.1, is violated. However, for the systems that we consider in the following chapters as non-positive recurrent, the former assumption implies the second one, see for instance, Remark 2.9. For this reason, by non-positive recurrence we will always assume that condition $\mathsf{E}T = \infty$ holds, in which case the subsequent asymptotic analysis does not depend on the specific value of T_1.

An important feature of a regenerative process $\{X(t)\}$ is that the regeneration property is preserved under an arbitrary measurable mapping f, implying that the process $\{f(X(t))\}$ is regenerative with the same *embedded renewal process* of regeneration instants $\{T_n\}$. (This stands in strong contrast to the *Markov property*, which can be lost after such a transformation.)

In this book, we only consider *non-negative* regenerative processes, and in order to formulate a few basic limit theorems below, it is convenient to represent a regenerative process as $\{f(X(t))\}$. Assume now that (1.2.4) holds and that the distribution of T is *non-lattice* (or *non-arithmetic*), i.e., not concentrated on the set $\{0, d, 2d, \ldots\}$ for any value of $d > 0$. If (1.2.4) holds, f is an indicator function, $f(X(t)) = \mathsf{1}(X(t) \in \cdot)$, and the weak limit (the limit in distribution) $X(t) \Rightarrow X$ exists, then the (stationary) distribution π of X satisfies [1, 3]

$$\pi(\cdot) = \lim_{t\to\infty} \mathsf{P}(X(t) \in \cdot) = \mathsf{P}(X \in \cdot) = \frac{\mathsf{E}\big(\int_0^T \mathsf{1}(X(t) \in \cdot)dt\big)}{\mathsf{E}T} . \qquad (1.2.5)$$

In this expression and similar relations below, the operator E denotes *expectation* with respect to the distribution of the process at a regeneration point corresponding to the zero-delayed case. (This distribution also appears in (2.6.2), where we denote it by φ.)

The result (1.2.5) demonstrates a fundamental importance of the notion of positive recurrence (1.2.4) in the regenerative stability analysis method. Therefore, in order to prove positive recurrence in the zero-delayed case, due to convergence (1.2.2), it suffices to show that there exists a deterministic (non-random) sequence $\{z_i\}$, $z_i \to \infty$, and constants $C \ge 0$, $\delta > 0$, such that the following *uniform* lower bound holds,

$$\inf_i \mathsf{P}(T(z_i) \le C) \ge \delta ,$$

implying $\mathsf{E}T < \infty$, while in the delayed case, we additionally must establish that $T_1 < \infty$ w.p.1. (Here, and in what follows, constants such as C and δ are always assumed to be finite.)

In the following chapters, we demonstrate that this approach turns out to be very effective for the purpose of finding sufficient stability conditions, and even stability criteria, of many queueing models. For this analysis, in Chap. 2, we also introduce the discrete-time version of a regenerative process, embedded at the arrival instants of the subsequent customers. This construction will also be intensively used in the performance analysis in the chapters that follow.

In the analysis presented in the subsequent chapters, processes with regenerative increments [3] play an important role as well.

Definition 1.3 A process $\{Y(t),\ t \ge 0\}$ is called a process with *regenerative increments* defined over the time instants $\{T_n,\ n \ge 0\}$, if the sequence $\{(T_{n+1} - T_n,\ \Delta_n)$, $n \ge 1\}$ is iid and independent of $(T_1,\ \Delta_0)$, where

$$T_0 := 0\ ,\quad \Delta_n := Y(T_{n+1}) - Y(T_n)\ ,\quad n \ge 0\,,$$

assuming $T_1 < \infty$, $\Delta_0 < \infty$ w.p.1.

Let us define the iid (cycle) maximums of the process $\{Y(t)\}$

$$M_n = \sup_{T_n \le t < T_{n+1}} |Y(t) - Y(T_n)|, \ n \ge 1, \tag{1.2.6}$$

with generic maximum M. In the following statement (which is a particular case of Theorem 54, Chap. 2 from [3]), Δ represents the generic (cycle) increment:

Theorem 1.1 *If* $\mathsf{E}\Delta < \infty$, $\mathsf{E}M < \infty$ *and* $\mathsf{E}T < \infty$, *then w.p.1,*

$$\lim_{t\to\infty} \frac{Y(t)}{t} = \frac{\mathsf{E}\Delta}{\mathsf{E}T}. \tag{1.2.7}$$

(Indeed, (1.2.7) holds if at least one of the mean values in the right-hand side is finite.)

In the remainder, we call the process with regenerative increments satisfying the assumptions of Theorem 1.1 *positive recurrent*. Next, we introduce the process

$$N(t) = \min(k \ge 1 : T_k > t), \tag{1.2.8}$$

which represents the number of regenerations in the time interval $[0, t]$, including the regeneration at instant 0, and which will be called a *renewal process*. The following form of *Wald's identity*, deduced from [3] (Proposition 53), is given for the zero-delayed process $\{Y(t)\}$ with regenerative increments (when $T_1 =_{st} T$):

$$\mathsf{E}Y(T_{N(t)}) = \mathsf{E}\sum_{n=1}^{N(t)} \Delta_n = \mathsf{E}N(t)\mathsf{E}\Delta, \ t \ge 0. \tag{1.2.9}$$

We also would like to highlight two classical results from renewal theory [3, 23]:

$$\lim_{t\to\infty} \frac{N(t)}{t} = \frac{1}{\mathsf{E}T} \quad \text{w.p.1}, \tag{1.2.10}$$

which is referred to as the SLLN for renewal processes, and the following *elementary renewal theorem*

$$\lim_{t\to\infty} \frac{\mathsf{E}N(t)}{t} = \frac{1}{\mathsf{E}T}. \tag{1.2.11}$$

(Both limits (1.2.10) and (1.2.11) equal zero if $\mathsf{E}T = \infty$.) If $\{f(X(t))\}$ is a regenerative process with regeneration instants $\{T_n\}$, then

$$Y(t) = \int_0^t f(X(u))du, \ t \ge 0, \tag{1.2.12}$$

is called the *cumulative process associated* with the regenerative process $\{f(X(t))\}$ [24]. Such a cumulative process is a particular case of the process with regenerative increments. In addition, the next result from [3] (Theorem 55, Chap. 2) is a special case of statement (1.2.7) for the cumulative process (1.2.12) (with non-negative mapping f):

Theorem 1.2 *Assume that the limiting distribution π of $f(X)$ in (1.2.5) exists, and that*

$$\mathsf{E}T < \infty\,,\ \mathsf{E}\Delta < \infty\,,\ \mathsf{E}f(X) = \int_0^\infty f(x)\pi(dx) < \infty\,. \tag{1.2.13}$$

Then the stationary performance measure $\mathsf{E}f(X)$ is obtained as the following w.p.1 limit:

$$\lim_{t\to\infty} \frac{1}{t}\int_0^t f(X((u))du = \frac{\mathsf{E}\Delta}{\mathsf{E}T} = \mathsf{E}f(X)\,. \tag{1.2.14}$$

The stationary measure $\mathsf{E}f(X)$ will often appear in the remainder as a byproduct of the stability analysis and here we present it separately as

$$\mathsf{E}f(X) = \frac{\mathsf{E}\int_0^T f(X(u))du}{\mathsf{E}T}\,, \tag{1.2.15}$$

where both sides can be infinite simultaneously, and $\int_0^T f(X(u))du = \Delta$.

Because a regenerative process $\{f(X(t))\}$ starts anew at each regeneration instant, then an important observation is that $\{f(X(t))\}$ can be treated as a particular case of the corresponding process with regenerative increments

$$\widetilde{\Delta}_n := f(X(T_{n+1})) - f(X(T_n))\,,\ n \ge 1\,,$$

with *zero mean* (generic) increment $\mathsf{E}\widetilde{\Delta} = 0$. Consequently, adopting the same notation M for the generic cycle maximum of $\{f(X(t))\}$, we arrive at the following statement (which is a particular case of Theorem 55, Chap. 2 from [3]):

Theorem 1.3 *If $\mathsf{E}T < \infty$, $\mathsf{E}M < \infty$, then, w.p.1,*

$$\lim_{t\to\infty} \frac{f(X(t))}{t} = \frac{\mathsf{E}\widetilde{\Delta}}{\mathsf{E}T} = 0\,. \tag{1.2.16}$$

Result (1.2.16) will be frequently used as well in the stability analysis presented in the following chapters. For instance, in the positive recurrent case, the remaining time process $\{T(t)\}$ satisfies

$$\lim_{t\to\infty} \frac{T(t)}{t} = 0\,,\ \ \text{w.p.1}\,, \tag{1.2.17}$$

since the iid increments

$$\Delta_n := T(T_{n+1}) - T(T_n) = T_{n+2} - T_{n+1} - (T_{n+1} - T_n),\ n \geq 1\ ,$$

have zero mean, and the (generic) maximum over a cycle satisfies the (stochastic) equality

$$\sup_{0 \leq t < T} T(t) =_{st} T\ .$$

For the reader's convenience, we summarize the most important asymptotic results considered in the subsequent chapters:

(i) The SLLN: if the (non-negative) summands $\{\xi_i\}$ are iid with $\mathsf{E}\xi < \infty$ then, w.p.1

$$\lim_{n\to\infty} \frac{1}{n} \sum_{i=1}^{n} \xi_i = \mathsf{E}\xi\ .$$

(ii) In case of *positive recurrence* of the involved processes, (1.2.14) holds if $\mathsf{E}\Delta \leq C\,\mathsf{E}T$ with a finite constant C. This property is valid, in particular, for the remaining work and queue-size processes, and also for an 'indicator'-type cumulative process $Y(t) = \int_0^t 1(f(X(u)) \in \cdot)du$ in (1.2.12) (such as the idle/busy time process).

(iii) The convergence in mean

$$\lim_{t\to\infty} \frac{1}{t} \mathsf{E} \int_0^t f(X(u))du = \mathsf{E}f(X)\,, \tag{1.2.18}$$

in general requires an extra moment assumption $\mathsf{E}[TM] < \infty$ (see [3]) which is sometimes hard to express in terms of the given variables. However, if the stationary distribution (1.2.5) exists, and if

$$\int_0^t f(X(u))du \leq Ct\ ,$$

for some constant C (again we refer to the idle/busy time process in this respect), then combined with the convergence w.p.1 in (1.2.14) (and hence *in probability*), the convergence in mean (1.2.18) follows from the *dominated convergence theorem* [3].

(iv) In what follows we often will deal with the *random sum* $\sum_{i=1}^{A(t)} \xi_i$, where the summation bound $A(t)$ represents the number of renewals (arrivals) in the interval $[0,\ t]$ and is either independent of the (non-negative) iid summands $\{\xi_i\}$ or is a *stopping time*, and with $\mathsf{E}\xi < \infty$. Then from Wald's identity and from the elementary renewal theorem (1.2.11), we obtain the following convergence in mean

$$\lim_{t\to\infty}\frac{1}{t}\mathsf{E}\sum_{i=1}^{A(t)}\xi_i = \lim_{t\to\infty}\frac{1}{t}\mathsf{E}A(t)\,\mathsf{E}\xi = \lambda\mathsf{E}\xi\,, \tag{1.2.19}$$

where λ is the rate of the renewal process.

Remark 1.2 Intuitively, a stopping time means that the 'decision' to stop and set $A(t) = n$ (or, more generally, $A(t) \geq n$) is based solely on the information contained in the first n summands, but not on any 'future' summand, see [3, 25, 26]. This property often (but not always!) holds in practice when dealing with a random summation.

1.3 Coupling

In this section, we briefly discuss some aspects of a special case of the notion of coupling of processes, leading to stochastic ordering, which is widely relied upon in the stability analyses presented throughout this book as well.

In general, the *coupling* of two processes $X,\ X'$ defined on a common state space but possibly with different probability spaces, is a realization $\widehat{X} = (\widehat{X},\ \widehat{X}')$ on a common probability space, such that the distributions of $X,\ X'$ are the marginals of the distribution of $\widehat{X}$. Let $X,\ Y$ be random variables with distribution functions $F_X,\ F_Y$. Then $X \leq_{st} (\geq_{st})\, Y$ in the sense of *stochastic ordering* if, for each value of x,

$$F_X(x) \geq (\leq)\, F_Y(x)\,,$$

or equivalently, there exist random variables $X',\ Y'$ (defined on a *common probability space*) such that

$$X' =_{st} X\,,\ Y' =_{st} Y\,,\quad \text{and}\ \ X' \leq (\geq)\, Y' \quad \text{w.p.1}\,.$$

These relations enable the comparison of processes in a sample-path sense. Stochastic ordering allows us to establish monotonicity properties of the involved queueing processes, which, in turn, play a critical role in the associated stability analysis by means of a construction that relies on *minorant* and/or *majorant* queueing systems with simpler (or known) stability conditions. The coupling method allows to replace random variables that have the same distribution functions, i.e., stochastically equivalent random variables, by variables that coincide w.p.1.

In this book, based on a predefined stochastic ordering of the governing random variables, on many occasions we first construct a coupling of the relevant processes, and then apply this to establish either the required bounds or the monotonicity of the output parameters. A clarifying example and application of this coupling technique can for instance be found in Sect. 3.3. In particular, the power of this approach is clearly demonstrated when we consider the system with non-identical servers in Sect. 4.4, and the retrial systems in Chaps. 7 and 8.

1.4 Notes

An elegant proof of property (1.2.3) (provided period T is non-lattice) can be found as a by-product of the analysis developed in [27] by means of a reduction of the problem to *Blackwell's renewal theorem* [3]. Renewal theory and the theory of regenerative and cumulative processes have mainly been developed in the famous works [22, 28]. The books [1, 3, 23] are also rich sources for results on modern renewal theory and related fields.

An important source of the monotonicity properties of queueing systems is [29], and the proofs of many basic results can be found in the papers [30–33]. Also see [7, 34] for more details on the coupling method.

The focus of this book is on the use of classical regenerative processes as defined above. However, we will briefly discuss several generalizations of regenerative processes in Sect. 10.4. (Note that the analysis related to the issue that is presented in this book has been briefly described in [35, 36] as well.)

The fundamental monographs [12, 14] are probably closest to the subject of present book, treating the stability of the Markov chains, and even more general classes of stochastic processes, on a high theoretical level, albeit with no focus on the stability of queueing systems. Another fundamental monograph [10], whose subject is very close to the one we study as well, contains a profound analysis of the stability of queueing processes in a highly general setting. Finally, one more outstanding book [9] is devoted to a profound theoretical analysis of the stability of queueing systems with ergodic and stationary input processes, and does not rely on the regenerative analysis method presented here.

References

1. Asmussen, S.: Applied Probability and Queues, 2nd edn. Springer, New York (2003)
2. Kalashnikov, V.: Topics on Regenerative Processes. CRC Press, Roca Baton (1994)
3. Serfozo, R.F.: Basics of Applied Stochastic Processes. Springer, Heidelberg (2009)
4. Shedler, G.S.: Regeneration and Networks of Queues. Springer, New York (1987)
5. Shedler, G.S.: Regenerative Stochastic Simulation. Academic Press Inc, San Diego (1993)
6. Sigman, K., Wolff, R.W.: A review of regenerative processes. SIAM Rev. **35**(2), 269–288 (1993)
7. Thorisson, H.: Coupling, Stationarity, and Regeneration, Probability and its Applications. Springer, New York (2000)
8. Feller, W.: An Introduction to Probability Theory and its Applications, II, 2nd edn. Wiley, New York (1971)

9. Baccelli, F., Bremaud, P.: Elements of Queueing Theory: Palm Martingale Calculus and Stochastic Recurrences, 2nd edn. Springer, New York (2003)
10. Borovkov, A.: Ergodicity and Stability of Stochastic Processes. Wiley, New York (2000)
11. Borovkov, A.A.: Asymptotic Methods in Queueing Theory. Wiley, New York (1984)
12. Borovkov, A.A.: Ergodicity and Stability of Stochastic Processes. Wiley Series in Probability and Statistics, 1st edn. (1998)
13. Chen, H., Yao, D.D. (eds.): Fundamentals of Queueing Networks: Performance, Asymptotics, and Optimization. Springer, New York (2001)
14. Meyn, S.P., Tweedie, R.L.: Markov Chains and Stochastic Stability. Springer, London (1993)
15. Robert, P.: Stochastic Networks and Queues. Springer, Berlin (2003)
16. Alfa, A.S.: Applied Discrete-Time Queues, 2nd edn. Springer, New York (2016)
17. Bhat, U.N.: An Introduction to Queueing Theory: Modeling and Analysis in Applications, 2nd edn. Birkhäuser Science, Basel (2015)
18. Bruneel, H., Kim, B.G.: Discrete-Time Models for Communication Systems Including ATM. Kluwer Academic Press, Boston (1993)
19. Cooper, R.B.: Introduction to Queueing Theory. North Holland, Amsterdam (1981)
20. Haribaskaran, G.: Probability. Queueing Theory and Reliability Engineering. Laxmi Publications, New Delhi (2005)
21. Kiefer, J., Wolfowitz, J.: On the theory of queues with many servers. Trans. Am. Math. Soc. **78**, 1–18 (1955)
22. Smith, W.L.: Regenerative stochastic processes. Proc. R. Soc. (Ser. A) **232**, 6–31 (1955)
23. Gut, A.: Stopped Random Walks Limit Theorems and Applications, 2nd edn. Springer Science+Business Media, Berlin (2009)
24. Glynn, P.W., Whitt, W.: Limit theorems for cumulative processes. Stoch. Proc. Appl. **47**(2), 299–314 (1993)
25. Billingsley, P.: Probability and Measure, 3rd edn. Wiley, New York (1995)
26. Shiryaev, A.: Probability. Springer, New York (1996)
27. Dynkin, E.B.: Some limit theorems for sums of independent random variables with infinite mathematical expectations. Sel. Trans. Math. Stat. Prob. **1**, 171–189 (1961)
28. Smith, W.L.: Renewal theory and its ramifications. J. Roy. Stat. Soc. (Series B) **20**(2), 243–302 (1958)
29. Müller, A., Stoyan, D.: Comparisons Methods for Stochastic Models and Risks. Wiley, Hoboken (2002)
30. Sonderman, D.: Comparing multi-server queues with finite waiting rooms, I: Same number of servers. Adv. Appl. Prob. **11**(2), 439–447 (1979)
31. Sonderman, D.: Comparing multi-server queues with finite waiting rooms, II: Different number of servers. Adv. Appl. Prob. **11**(2), 448–455 (1979)
32. Whitt, W.: Comparing counting processes and queues. Adv. Appl. Prob. **13**, 207–220 (1981)
33. Whitt, W.: Blocking when service is required from several facilities simultaneously. AT&T Tech. J. **64**(8), 1807–1856 (1985)
34. Lindvall, T.: Lectures on the Coupling Method. Wiley, New York (1992)
35. Morozov, E., Delgado, R.: Stability analysis of regenerative queues. Autom. Rem. Contr. **70**, 1977–1991 (2009)
36. Morozov, E., Steyaert, B.: A stability analysis method of regenerative queueing systems. In: Anisimov, V., Limnios, N. (eds.) Queueing Theory 2, Advanced Trends, 239–268. Wiley/ISTE, New York (2021)

Chapter 2
The Classical $GI/G/1$ and $GI/G/m$ Queueing Systems

In this chapter, we demonstrate the main steps of the regenerative stability analysis by considering the classical single-server $GI/G/1$ queueing system, as well as the multiserver $GI/G/m$ system with $m > 1$ servers.

2.1 The Single-Server System

2.1.1 Description of the Model

We consider a single-server $GI/G/1$ queueing system with *infinite buffer* depicted in Fig. 2.1. The customer arrival process is a general renewal process with arrival instants $\{t_n, n \geq 1\}$ and iid interarrival times $\{\tau_n = t_{n+1} - t_n\}$ with arrival rate $\lambda = 1/\mathsf{E}\tau$. In what follows, the $1st$ customer is assumed to arrive at instant $t_1 = 0$ (this assumption can be relaxed if multiple input processes are considered). In addition, let $\{S_n, n \geq 1\}$ be the sequence of iid customer service times with *service rate* $\mu = 1/\mathsf{E}S$.

As mentioned before, here and in the following chapters, we omit the sequential index to denote a generic element of an iid sequence of random variables (such as τ and S). Denote by $W(t)$ and $\nu(t)$ the workload process (i.e., the remaining amount of work) and the number of customers in the system respectively, at time instant t. We call these the *basic processes*, and adopt the following notation for the processes

$$W(t_k^-) = W_k, \quad \nu(t_k^-) = \nu_k, \;\; k \geq 1, \tag{2.1.1}$$

embedded into the basic processes at customer arrival instants. To be more precise, the embedded processes are considered *just before arrivals*, which corresponds to the system state observed by newly arriving customers.

E. Morozov and B. Steyaert, *Stability Analysis of Regenerative Queueing Models*,
https://doi.org/10.1007/978-3-030-82438-9_2

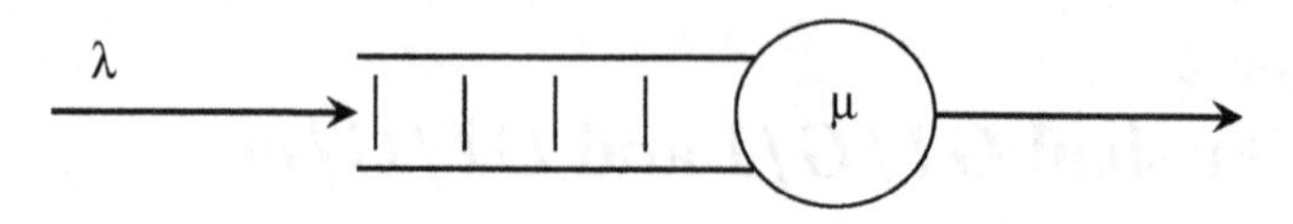

Fig. 2.1 The $GI/G/1$ single-server queueing system

Also, in case of a *zero initial state*, the 1*st* customer encounters an empty system upon arrival, and hence in this case $W_1 = \nu_1 = 0$, while in the case of an *arbitrary initial state*, in general, $W_1,\ \nu_1 \in [0,\ \infty)$ w.p.1. We initialize by setting $T_0 = 0$, and the regeneration instants of the continuous-time processes $\{\nu(t)\}$ and $\{W(t)\}$ are then defined as

$$T_{n+1} = \min_{k\geq 1}\,(t_k > T_n : \nu_k = W_k = 0)\ , \quad n \geq 0\,. \tag{2.1.2}$$

Therefore, at each time instant T_n, the arriving customer observes the system completely idle and, as it is easy to see, the basic processes start anew independently of the prehistory, which corresponds to the construction of a classically regenerative process in Sect. 1.2. We return to this construction in Remark 2.1 below. We further define

$$\theta_{n+1} = \min_{k\geq 1}\,(k > \theta_n : \nu_k = W_k = 0)\ , \quad n \geq 0\,, \quad \theta_0 := 1\,, \tag{2.1.3}$$

which (for $n \geq 1$) are the regeneration instants of the (embedded) sequences $\{\nu_k\}$ and $\{W_k\}$. (Note that $\theta_1 > 1$.) In particular,

$$T_1 = \tau_1 + \cdots + \tau_{\theta_1 - 1}\ ; \quad T_n = t_{\theta_n}\ , \quad n \geq 0\,. \tag{2.1.4}$$

As in Sect. 1.2, let T represent a generic regeneration period in continuous time, and denote the generic regeneration period in discrete time by θ. The generic period θ equals the number of arrivals in the queueing system during the continuous-time regeneration period T. If $\nu_1 = 0$, then the corresponding queueing process is called *zero-delayed*, in which case $T_1 =_{st} T,\ \theta_1 - 1 =_{st} \theta$. Note that the generic periods T and θ are related by the stochastic equality

$$T =_{st} \tau_1 + \cdots + \tau_\theta\ , \tag{2.1.5}$$

and, using Wald's identity, we obtain

$$\mathsf{E}T = \mathsf{E}\tau\ \mathsf{E}\theta\,. \tag{2.1.6}$$

In view of (2.1.6), $\mathsf{E}\,\theta < \infty$ holds if and only if $\mathsf{E}\,T < \infty$. The sequence $\{T_n\}$ is called *positive recurrent* if (1.2.4) holds. Similarly, we call the sequence $\{\theta_n\}$ positive recurrent if

$$\theta_1 < \infty \quad \text{w.p.1 and} \quad \mathsf{E}\,\theta < \infty\,.$$

Under a general (non-zero) initial state, we refer to the queueing process as *delayed*. Note that due to relations (2.1.4), $T_1 < \infty$ holds if and only if $\theta_1 < \infty$, because $\tau < \infty$ w.p.1. Therefore, it suffices to prove the positive recurrence of either the continuous-time or the discrete-time processes.

In short, we say that a queueing system with dynamics described by a regenerative process is *positive recurrent* if the renewal process of regenerations is so, see Definition 1.2.

For a discrete-time regenerative process, we define the remaining regeneration time at instant n as

$$\theta(n) = \min_{k\geq 1}\,(\,\theta_k - n : \theta_k > n)\,,\ n \geq 1\,.$$

Then in particular, the length of the $1st$ regeneration period $\theta(1) = \theta_1 - 1$, and, in the zero-delayed case, $\theta(1) =_{st} \theta$. Moreover, if $\mathsf{E}\theta = \infty$, then $\theta(n) \Rightarrow \infty$, or equivalently, for each $k \geq 1$ (see (1.2.3))

$$\mathsf{P}(\theta(n) > k) \to 1\,,\ n \to \infty\,.$$

Note that θ is called *periodic* with period $d > 1$ if

$$\mathsf{P}\Big(\bigcup_{n=1}^{\infty}\{\theta = nd\}\Big) = 1\,,$$

and θ is *aperiodic* if no such integer d exists. If θ is aperiodic, then in the positive recurrent case, $X_n \Rightarrow X$, the stationary distribution of X is defined as

$$\lim_{n\to\infty} \mathsf{P}(X_n \in \cdot) = \mathsf{P}(X \in \cdot) = \frac{\mathsf{E}\sum_{k=1}^{\theta} \mathsf{1}(X_k \in \cdot)}{\mathsf{E}\theta}\,,$$

and the corresponding stationary performance measure $\mathsf{E}f(X)$ is obtained as

$$\lim_{n\to\infty} \frac{1}{n}\sum_{k=1}^{n} f(X_k) = \frac{\mathsf{E}\sum_{k=1}^{\theta} f(X_k)}{\mathsf{E}\,\theta} = \mathsf{E}f(X)\,, \tag{2.1.7}$$

under the appropriate assumptions, see (1.2.13)–(1.2.15).

Remark 2.1 At first glance, it seems possible to take as the start of a new regeneration cycle, for instance, the time instant t_0 of the beginning of an idle period, when $W(t)$ and $\nu(t)$ become zero for the first time in the current regeneration cycle. However, this is not a correct approach, because the future dynamics of the queueing processes after instant t_0 in general depend on the remaining (interarrival) time up to the next arrival, which in turn depends on the attained interarrival time. (This effect can be observed on Fig. 2.8 in Sect. 2.5.2.) As a result, consecutive regeneration cycles would no longer be independent in such a setting. Nevertheless, when the input process is Poisson, then such a choice for a regeneration point is possible (as well as any other time instant within the idle period) because, due to the memoryless property, the remaining interarrival time is distributed as the original exponential interarrival time τ. In this case, in order to construct a regeneration point, we can replace the remaining interarrival time by an *independent copy* of τ. This replacement leaves the distribution of all involved processes unchanged.

2.1.2 *Stability Analysis*

For the $GI/G/1$ queueing system described above, its stability (positive recurrence) is determined by the *traffic intensity*

$$\rho := \frac{\mathsf{E}S}{\mathsf{E}\tau} = \frac{\lambda}{\mu}.$$

Theorem 2.1 *If $\rho < 1$, then*
(i) the system is positive recurrent for any initial state;
(ii) W_n, ν_n have stationary distributions as $n \to \infty$;
(iii) $W(t)$, $\nu(t)$ have stationary distributions as $t \to \infty$ if τ is non-lattice.
If the system is positive recurrent, then $\rho < 1$.

Proof (i) We introduce the idle time of the server

$$I(t) = \int_0^t 1(\nu(u) = 0)du ,$$

and the number of arrivals $A(t)$ in the interval $[0, t]$. Note that $A(t_n) = n, \ n \geq 1$. Then the total workload that has arrived in $[0, t]$ is given by

$$V(t) = \sum_{n=1}^{A(t)} S_n, \ \ t \geq 0 .$$

Obviously,

$$W_1 + V(t) = t - I(t) + W(t) \geq t - I(t) , \tag{2.1.8}$$

and hence, the following inequality holds:

$$I(t) \geq t - V(t) - W_1, \ \ t \geq 0 .$$

Invoking the SLLN, we obtain (w.p.1),

$$\lim_{t\to\infty}\frac{V(t)}{t}=\lim_{t\to\infty}\frac{A(t)}{t}\frac{\sum_{n=1}^{A(t)}S_n}{A(t)}=\lambda\,\mathsf{E}S=\rho\,, \tag{2.1.9}$$

implying

$$\liminf_{t\to\infty}\frac{I(t)}{t}\geq 1-\rho>0\,. \tag{2.1.10}$$

In particular, the idle time satisfies

$$I(t)\to\infty\quad \text{w. p. 1 as } t\to\infty\,. \tag{2.1.11}$$

Because of $I(t)/t\leq 1,\ t>0$, then it follows from (2.1.10), *Fatou's lemma* and *Fubini's theorem* [1] (allowing to interchange the expectation and integral operators) that

$$\liminf_{t\to\infty}\frac{\mathsf{E}\,I(t)}{t}=\liminf_{t\to\infty}\frac{1}{t}\int_0^t\mathsf{P}(\nu(u)=0)du\geq\mathsf{E}\liminf_{t\to\infty}\frac{I(t)}{t}\geq 1-\rho\,. \tag{2.1.12}$$

It is easy to check that (2.1.12) implies $\mathsf{P}(\nu(t)=0)\not\to 0$, and therefore there exists a deterministic (*non-random*) sequence of time instants $z_i\to\infty$ and a constant $\delta>0$ such that

$$\inf_i\ \mathsf{P}(\nu(z_i)=0)\geq\delta\,. \tag{2.1.13}$$

Define the *remaining interarrival time* at instant t as

$$\tau(t)=\min_{k\geq 1}\,(t_k-t:\,t_k>t),\quad t\geq 0\,. \tag{2.1.14}$$

(Note that $\tau(t_k^-)=0,\ \tau(t_k)=\tau_k,\ k\geq 1$.) As it is shown in Chap. 3 (see (3.1.4)), the process $\{\tau(t)\}$ is *tight*, and therefore, there exists a constant C such that

$$\inf_{t\geq 0}\ \mathsf{P}(\tau(t)\leq C)\geq 1-\delta/2\,, \tag{2.1.15}$$

where δ satisfies (2.1.13). Since the instants $\{z_i\}$ are deterministic, then (2.1.15) holds as well when the parameter t is replaced by $\{z_i\}$. Note that, by conditioning on the event $\{\nu(z_i)=0\}$, the first arrival after instant z_i initiates a regeneration of the process $\{\nu(t)\}$. As a result, we find from (2.1.13) and (2.1.15) that, for an arbitrary z_i,

$$\begin{aligned}\mathsf{P}(T(z_i)\leq C)&\geq\ \mathsf{P}\big(\nu(z_i)=0,\ \tau(z_i)\leq C\big)\\&\geq\ \mathsf{P}(\nu(z_i)=0)-\ \mathsf{P}(\tau(z_i)>C)\geq\frac{\delta}{2}\,.\end{aligned} \tag{2.1.16}$$

Since the values of C and δ in (2.1.16) do not depend on z_i and i, it then follows that $T(t) \nRightarrow \infty$ indeed holds. This implies $\mathsf{E}T < \infty$ and $\mathsf{E}\theta < \infty$ in view of (1.2.2) and (2.1.6).

In the delayed case, due to (2.1.11), the initial busy period u_0 is finite:

$$u_0 := \inf(t > 0 : W(t) = 0) = \inf(t > 0 : I(t) > 0) < \infty \quad \text{w.p.1} .$$

Because the $1st$ regeneration (after instant $t = 0$) occurs at instant $u_0 + \tau(u_0)$, then

$$T_1 = u_0 + \tau(u_0) < \infty \quad \text{w.p.1} ; \quad \theta_1 = A(T_1) < \infty \quad \text{w.p.1} .$$

(ii) Using assumption $\rho < 1$ written as $\mathsf{E}\tau > \mathsf{E}S$, we obtain $\mathsf{E}(\tau - S) > 0$, implying $\mathsf{P}(\tau - S > 0) > 0$, and therefore

$$\mathsf{P}(\theta = 1) = \mathsf{P}(\tau > S) > 0 , \tag{2.1.17}$$

i.e., the regeneration period θ is aperiodic. As a result, W_n and ν_n have stationary distributions as $n \to \infty$.

(iii) If τ is non-lattice, then period T is non-lattice as well (see (2.1.5)), and by (1.2.5), $W(t)$, $\nu(t)$ have stationary distributions as $t \to \infty$ (see Chapter VI in [2]).

Assume that the system is positive recurrent. By (2.1.17),

$$\mathsf{P}(\tau \geq S + \delta_0) \geq \varepsilon_0 , \tag{2.1.18}$$

for some constants $\varepsilon_0, \ \delta_0 > 0$. Condition (2.1.18) shows that if a customer arrives in an empty system, then, with a probability which is no less than ε_0, the system will be idle during no less than the time δ_0 before the next customer arrives. (This scenario is illustrated in Fig. 2.2.) It follows from *equality* in (2.1.8) and from (2.1.9) that

$$\lim_{t\to\infty} \frac{V(t)}{t} \equiv \rho = 1 - \frac{\mathsf{E}I_0}{\mathsf{E}T} , \tag{2.1.19}$$

where I_0 is the idle period of the server during a regeneration period, and we take into account that $W(t) = o(t)$ w.p.1, see (2.3.22) below. Now we obtain from (2.1.18) that the mean idle period satisfies

$$\mathsf{E}I_0 \geq \mathsf{E}(I_0; I_0 \geq \delta_0) = \mathsf{E}(I_0 \,|\, \tau \geq S + \delta_0)\, \mathsf{P}(\tau \geq S + \delta_0) \geq \varepsilon_0\delta_0 > 0 . \tag{2.1.20}$$

Then (2.1.18), (2.1.20) imply condition $\rho < 1$, which is indeed the stability criterion. □

Problem 2.1 *Show that* $\max_{n\geq 1} \mathsf{E}W_n < \infty$ *implies* $W_n \nRightarrow \infty$.

Problem 2.2 *Explain the equality in (2.1.17).*

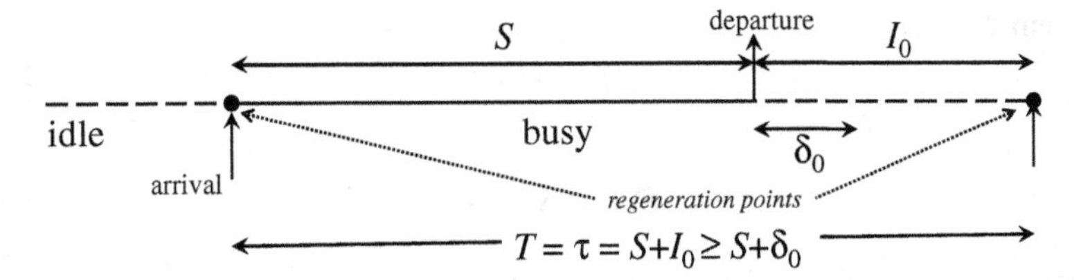

Fig. 2.2 A regeneration cycle in a single-server system containing only 1 customer, with idle period $I_0 \geq \delta_0$

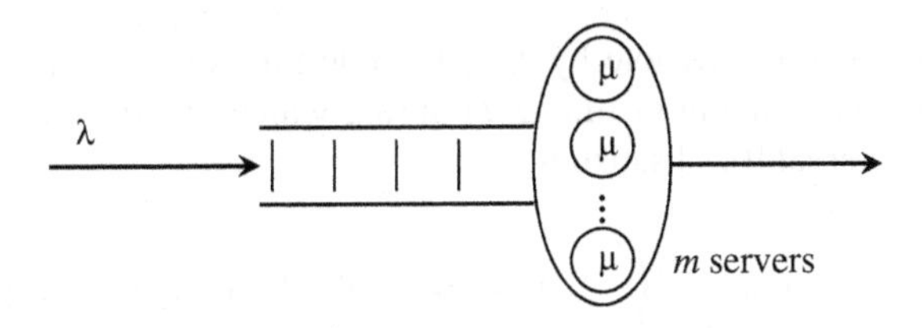

Fig. 2.3 The $GI/G/m$ multiserver queueing system

2.2 The Multiserver System

Consider a FCFS multiserver $GI/G/m$ queueing system with identical servers and infinite buffer, as depicted in Fig. 2.3, with the same renewal input process with rate λ as above, iid service times $\{S_n\}$ with rate $\mu = 1/\,\mathsf{E}S$, and traffic intensity $\rho = \lambda\,\mathsf{E}S$. Denote by $W_i(t)$ the remaining work in server i at instant t, and let

$$\mathsf{W}(t) = \sum_{i=1}^{m} W_i(t),\ \ t \geq 0\,, \tag{2.2.1}$$

be the aggregated remaining work. Consider the process

$$\mathbf{Y}(t) = \big(\nu(t),\ \mathsf{W}(t),\ \tau(t)\big)\ ,\ t \geq 0\,.$$

This process regenerates at the instants $\{T_n\}$ defined as

$$T_{n+1} = \min_k(t_k > T_n : \mathbf{Y}_k = \mathbf{0}),\ \ n \geq 0 \quad (T_0 := 0)\,,$$

where

$$\mathbf{Y}_k := \mathbf{Y}(t_k^-) \equiv (\nu_k,\ \mathsf{W}_k,\ 0)\,,$$

and the equality $\mathbf{Y}_k = \mathbf{0}$ is component-wise. (Recall that $\tau(t_k^-) = 0$ due to definition (2.1.14).)

Theorem 2.2 *(i) If*

$$\rho < m\,, \tag{2.2.2}$$

and

$$\mathsf{P}(\tau > S) > 0\,, \tag{2.2.3}$$

then $\mathsf{E}T < \infty$*, i.e., the zero-delayed system is positive recurrent. (ii) If the system is positive recurrent, i.e.,* $T_1 < \infty$, $\mathsf{E}T < \infty$*, then both conditions (2.2.2) and (2.2.3) hold.*

Proof First we prove (i). Denote by $I_i(t)$ the idle time of server i in the interval $[0,\ t]$. Then the total amount of time $I(t)$ during which *at least one server is idle* within the time interval $[0,\ t]$ satisfies,

$$I(t) = \int_0^t \mathsf{1}(\nu(u) < m)du \geq \int_0^t \mathsf{1}(W_i(u) = 0)du = I_i(t)\,,\quad i = 1, \ldots, m,\quad t \geq 0\,.$$

Hence,

$$mI(t) \geq \sum_{i=1}^{m} I_i(t) =: \widehat{I}(t)\,, \tag{2.2.4}$$

and similar to the single-server case, we can write down the following relation

$$\mathsf{W}_1 + V(t) = tm - \widehat{I}(t) + \mathsf{W}(t) \geq m(t - I(t))\,,\quad t \geq 0\,. \tag{2.2.5}$$

In view of $\rho < m$, this implies as above (see (2.1.10) and (2.1.12))

$$\liminf_{t\to\infty} \frac{I(t)}{t} \geq 1 - \frac{\rho}{m} =: \delta_0 > 0\,, \tag{2.2.6}$$

and

$$\liminf_{t\to\infty} \frac{\mathsf{E}I(t)}{t} = \liminf_{t\to\infty} \frac{1}{t}\int_0^t \mathsf{P}(\nu(u) < m)du \geq \delta_0\,.$$

Then the inequality

$$\limsup_{t\to\infty} \mathsf{P}(\nu(t) < m) > 0\,,$$

holds, and consequently,

$$\inf_i \mathsf{P}(\nu(z_i) < m) \geq \delta\,, \tag{2.2.7}$$

holds as well for a deterministic sequence $z_i \to \infty$ and some constant $\delta > 0$. Next, we introduce the *remaining service time* $S_i(t)$ in server i at instant t (we set $S_i(t) = 0$ if the server is idle). In Chap. 3, we prove that the processes $\{\tau(t)\}$ and $\{S_i(t)\}$ are *tight*, $i = 1, \ldots, m$, see Theorem 3.3, and (3.1.4). Consequently, the aggregated process $\{\sum_{i=1}^m S_i(t),\ t \geq 0\}$ is tight as well. For any constant $D > 0$ and any instant z_i satisfying (2.2.7), we define the set

$$\mathbb{B}_D = [0,\ m-1] \times [0,\ D] \times [0,\ D]\,,$$

and the event

$$\mathcal{E}_i(D) = \Big\{\mathbf{Y}(z_i) \in \mathbb{B}_D\Big\}\,.$$

Note that $\mathsf{W}(z_i) = \sum_{k=1}^m S_k(z_i)$, provided that $\nu(z_i) < m$. Then it is easy to deduce that

$$\begin{aligned}
\mathsf{P}(\mathcal{E}_i(D)) \geq\ & \mathsf{P}(\nu(z_i) < m) \\
& - \mathsf{P}(\sum_{k=1}^m S_k(z_i) > D) - \mathsf{P}(\tau(z_i) > D) \\
& - \mathsf{P}(\sum_{k=1}^m S_k(z_i) > D,\ \tau(z_i) > D) \\
\geq\ & \mathsf{P}(\nu(z_i) < m) - \mathsf{P}(\sum_{k=1}^m S_k(z_i) > D) - 2\,\mathsf{P}(\tau(z_i) > D).
\end{aligned} \tag{2.2.8}$$

Due the tightness, we can set the constant D such that

$$\begin{aligned}
\mathsf{P}\big(\sum_{k=1}^m S_k(z_i) > D\big) &\leq \frac{\delta}{8}\,, \\
\mathsf{P}(\tau(z_i) > D) &\leq \frac{\delta}{8}\,,
\end{aligned} \tag{2.2.9}$$

where δ satisfies (2.2.7). Then (2.2.7)–(2.2.9) imply

$$\mathsf{P}(\mathcal{E}_i(D)) = \mathsf{P}\Big(\nu(z_i) < m,\ \mathsf{W}(z_i) \leq D,\ \tau(z_i) \leq D\Big) \geq \delta/2\,. \tag{2.2.10}$$

Relying on $\mathsf{P}(\tau > S) > 0$ and due to $\mathsf{E}\tau < \infty$, there exist positive constants $\varepsilon_0,\ \delta_1,\ C$ such that

$$\mathsf{P}(C \geq \tau > S + \varepsilon_0) \geq \delta_1\,. \tag{2.2.11}$$

Define now the integer $L = \lceil D/\varepsilon_0 \rceil$ and, for a fixed (arbitrary) instant z_i, denote by

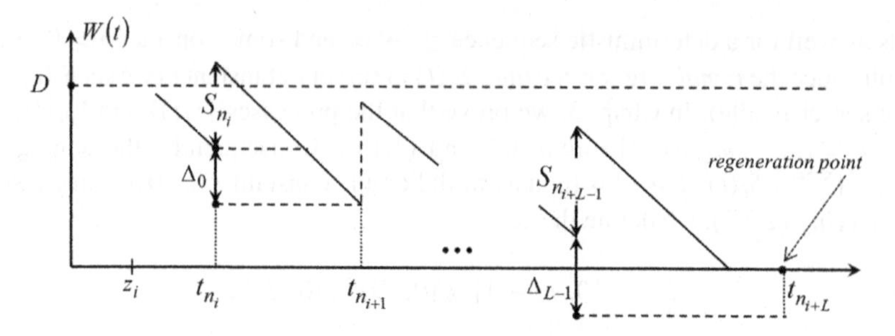

Fig. 2.4 Unloading the system by L arrivals: within the k-th interarrival time the workload process decreases, according to assumption (2.2.11), by no less than ε_0

$$n_i = \min(k \geq 1 : t_k > z_i)\,,$$

the index of the first arrival after z_i. Also introduce the event

$$\mathcal{A}_i(L) = \bigcap_{k=0}^{L-1} \left\{ C \geq \tau_{n_i+k} > S_{n_i+k} + \varepsilon_0 \right\}, \tag{2.2.12}$$

which is composed of independent individual events. The event (2.2.12) implies that (starting at time instant t_{n_i}) every interarrival time is upper bounded by the constant C and exceeds the service time of the new customer by no less than ε_0. As a result, during each interarrival time, the workload process decreases by no less than ε_0 as long as the system is not completely empty. This procedure is shown in Fig. 2.4, where (for a fixed z_i) we denote by

$$\Delta_k = \tau_{n_i+k} - S_{n_i+k}\,, \quad k = 0, \ldots, L-1\,,$$

the values of the decrement of the workload process within sequential interarrival times during the unloading procedure. Hence, we can expect that at most L such events lead to an empty state of the system, and a new regeneration cycle starts.

Indeed, conditioned on the event $\mathcal{E}_i(D) \cap \mathcal{A}_i(L)$, a customer that observes an empty system arrives in the time interval $[z_i, z_i + D + LC]$, and consequently, a regeneration occurs in this interval. Since by (2.2.11)

$$\mathsf{P}(\mathcal{A}_i(L)) \geq \delta_1^L\,,$$

and the events $\mathcal{E}_i(D)$ and $\mathcal{A}_i(L)$ are independent, then, also invoking (2.2.10), the following lower bound holds,

$$\mathsf{P}\big(T(z_i) \leq D + LC\big) \geq \mathsf{P}\big(\mathcal{E}_i(D) \cap \mathcal{A}_i(L)\big) \geq \frac{\delta}{2}\,\delta_1^L > 0\,, \tag{2.2.13}$$

which is uniform in z_i and i. This implies $T(t) \not\Rightarrow \infty$, and hence $\mathsf{E}T < \infty$, which concludes the proof. Strictly speaking, in a detailed calculation of $T(z_i)$ we must also take into account the distance $t_{n_i} - z_i = \tau(z_i)$ up to the $1st$ arrival after instant z_i. Since the family $\{\tau(z_i)\}$ is tight, we can choose a large enough constant D_0 such that $\tau(z_i) \le D_0$ and the uniform lower bound still holds:

$$\inf_i \mathsf{P}\big(T(z_i) \le D_0 + D + LC\big) > 0.$$

(ii) We first rewrite equality (2.2.5) as

$$\widehat{I}(t) = tm + \mathsf{W}(t) - \mathsf{W}_1 - V(t)\,. \qquad (2.2.14)$$

In (2.3.22) below, we show that $\mathsf{W}(t) = o(t)$ (also see (1.2.17)). Then, by analogy with (2.1.19), it follows from (2.2.14) that

$$\lim_{t\to\infty} \frac{\widehat{I}(t)}{t} = \frac{\mathsf{E}\widehat{I}_0}{\mathsf{E}T} = \rho - m\,, \qquad (2.2.15)$$

where $\widehat{I}_0$ is the increment of the process $\{\widehat{I}(t)\}$ during a regeneration period. If assumption $\mathsf{P}(\tau > S) > 0$ holds, then condition $\rho < m$ immediately follows in the limit from (2.1.20), (2.2.14) and (2.2.15).

Nonetheless, assumption $\mathsf{P}(\tau > S) > 0$ is indeed superfluous and follows from the positive recurrence assumption. In order to prove this, denote by P_0 the stationary probability that the system is *completely idle*. Then

$$\lim_{t\to\infty} \frac{1}{t} \int_0^t \mathbb{1}(\nu(u) = 0)du = \mathsf{P}_0 = \frac{\mathsf{E}I_0^{(0)}}{\mathsf{E}T}\,, \qquad (2.2.16)$$

where $I_0^{(0)}$ is the length of the idle period preceding a regeneration point, when *all servers are simultaneously idle*. (Note that $I_0^{(0)} = I_0$ in the single-server system.) Assume for a moment that

$$\mathsf{P}(\tau \le S) = 1\,, \qquad (2.2.17)$$

and we show that then classical regenerations do not exist. Denote by $D(t)$ the number of departures within the interval $[0,\ t]$. It is sufficient to consider the zero initial state single-server system. First, assume that $\mathsf{P}(\tau < S) = 1$, then evidently (classical) regenerations are impossible. If $\mathsf{P}(\tau = S) > 0$, then it is easy to see that the number of customers in the system is lower bounded by 1:

$$\nu(t) = A(t) - D(t) \ge k - (k-1) = 1,\ t \in [t_k,\ t_{k+1}),\ k \ge 1\,, \qquad (2.2.18)$$

i.e., regenerations are impossible as well. Therefore, under positive recurrence, condition $\mathsf{P}(\tau > S) > 0$ is valid and $\mathsf{E}I_0^{(0)} > 0$ (see (2.1.20)), and then (2.2.16) implies $\mathsf{P}_0 > 0$. Because $I_0^{(0)} \le \widehat{I}_0$, then condition $\rho < m$ follows from (2.2.15). □

Remark 2.2 We repeat the definitions

$$\mathbf{Y}(t) = (\nu(t),\, \mathsf{W}(t),\, \tau(t)),\;\; \mathbf{Y}_n = (\nu_n,\, \mathsf{W}_n,\, 0)\,,$$

and point out that, since $z_i < t_{n_i}$, then $\mathbf{Y}(z_i) \ge \mathbf{Y}_{n_i}$ (in the component-wise sense) and

$$\{\mathbf{Y}(z_i) \in \mathbb{B}_D\} \subseteq \{\mathbf{Y}_{n_i} \in \mathbb{B}_D\}\,.$$

Then, under assumptions (2.2.2), (2.2.3), $\mathsf{P}(\mathbf{Y}_{n_i} \in \mathbb{B}_D) \ge \delta/2$ due to (2.2.10). Since, conditioned on the event $\mathcal{E}_i(D)$, the event $\mathcal{A}_i(L)$ implies a regeneration after at most L arrivals, then the (discrete) remaining regeneration time $\theta(n_i)$ is upper bounded uniformly in i as in (2.2.13), implying that

$$\mathsf{P}(\theta(n_i) \le L) \ge \frac{\delta}{2}\,\delta_1^L\,, \tag{2.2.19}$$

leading to $\mathsf{E}\theta < \infty,\;\; \mathsf{E}T < \infty$.

2.3 The Finiteness of the First Regeneration Period

Now we establish the finiteness (w.p.1) of the first regeneration period, using an approach from [3]. We split the proof into two steps. First, we show that the number of customer arrivals that observe the queueing process in a bounded set containing a regeneration state, increases unlimitedly as $t \to \infty$. Secondly, we show that, within the first regeneration cycle, the number of arrivals observing the basic process in each such a set is finite w.p.1. It then follows that the number of regeneration cycles is no less than two and therefore $\theta_1 < \infty$, $T_1 < \infty$ w.p.1.

In addition to the original processes $\{\mathbf{Y}(t)\}$ and $\{\mathbf{Y}_n\}$, in the proof of Theorem 2.3 below, we will also make use of the following processes

$$\mathbf{Z}(t) = (\nu(t),\, \mathbf{S}(t)),\; t \ge 0\,;\quad \mathbf{Z}_n = (\nu_n,\, \mathbf{S}(t_n^-))\;,\; n \ge 1\,,$$

where $\mathbf{S}(t) = (S_1(t), \ldots, S_m(t))$ is the vector of the remaining service times at time t.

Remark 2.3 Evidently, the processes $\{\mathbf{Z}(t)\}$ and $\{\mathbf{Z}_n\}$ regenerate at the time instants (2.1.2) and (2.1.3), respectively. Actually, depending on the purpose, we may consider any suitable process with regenerations (2.1.2) or (2.1.3).

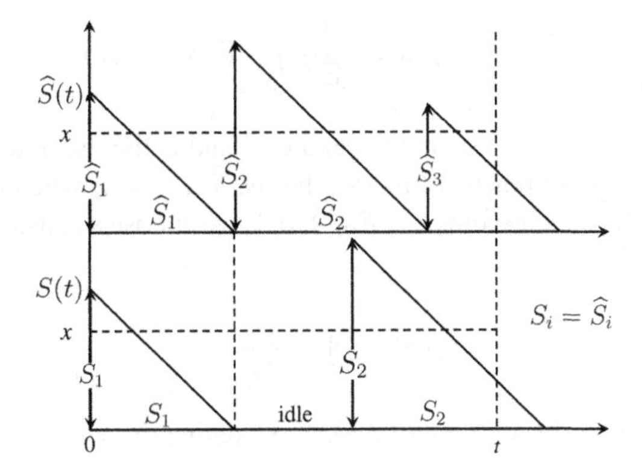

Fig. 2.5 Illustration of inequality (2.3.2)

Theorem 2.3 *If* $\rho < m$ *and* $\mathsf{P}(\tau > S) > 0$, *then the process* $\{\mathbf{Y}(t)\}$ *is positive recurrent for any initial state* $\mathbf{Y}_1$. *In particular,*

$$\theta_1 < \infty , \quad T_1 < \infty \quad w.p.1 . \tag{2.3.1}$$

Proof In the light of Theorem 2.2, it is sufficient to merely prove (2.3.1).

Step 1. We consider the iid service times sequences $\{S_n^{(i)}\}$ which are also independent for different values of i and are copies of a generic service time S. For each fixed i, consider a *delayed renewal process*

$$\widehat{Z}_0^{(i)} := S_i(0) , \ \widehat{Z}_n^{(i)} := S_i(0) + S_1^{(i)} + \cdots + S_n^{(i)}, \ n \geq 1 ,$$

where the initial intervals $S_i(0)$ and service times $\{S_n^{(i)}\}$ are *those that occur in the real service process* in server i, and define the remaining renewal time at instant t as

$$\widehat{S}_i(t) = \min_{n \geq 0}(\widehat{Z}_n^{(i)} - t : \widehat{Z}_n^{(i)} > t), \ t \geq 0 .$$

Although the actual remaining service time $S_i(t)$ and the remaining renewal time $\widehat{S}_i(t)$ in general are not comparable due to the possible *empty periods* of the servers, for an arbitrary value $x \geq 0$ we still obtain the following inequalities illustrated by Fig. 2.5:

$$\int_0^t \mathsf{1}(S_i(u) \leq x)du \geq \int_0^t \mathsf{1}(\widehat{S}_i(u) \leq x)du , \quad t \geq 0 , \ i = 1, \ldots, m . \tag{2.3.2}$$

For each x, the process $\{1(\widehat{S}_i(t) > x)\}$ is regenerative with generic regeneration period S, and it then follows from (1.2.5), (1.2.14) that (see Problem 2.7 below)

$$\lim_{t\to\infty}\frac{1}{t}\int_0^t 1(\widehat{S}_i(u)\le x)du = \frac{1}{\mathsf{E}S}\int_0^x \mathsf{P}(S>u)du =: F_e(x)\,, \quad (2.3.3)$$

where $F_e(x) = \mathsf{P}(S_e \le x)$ does not depend on i and is the distribution function of the *stationary* remaining renewal time S_e, that is, $\widehat{S}_i(t) \Rightarrow S_e$ when the limit exists. (Also see Sect. 4.5.) Because $\lim_{x\to\infty} F_e(x) = 1$, we choose a constant $M \in (0,\, \infty)$ such that

$$F_e(M) \ge 1 - \frac{\delta_0}{2m}\,, \quad (2.3.4)$$

with $\delta_0 = 1 - \rho/m > 0$ satisfying (2.2.6). Next, define the set

$$\mathbb{B}_M = [0,\, M] \times \cdots \times [0,\, M] \in \mathbb{R}_+^m\,.$$

Then it easily follows from (2.3.2)–(2.3.4) that

$$\begin{aligned}
&\liminf_{t\to\infty}\frac{1}{t}\int_0^t 1(\mathbf{S}(u)\in\mathbb{B}_M)du \ge \\
&\liminf_{t\to\infty}\frac{1}{t}\int_0^t 1(S_1(u)\le M)du - \limsup_{t\to\infty}\frac{1}{t}\sum_{i=2}^m\int_0^t 1(S_i(u)>M)du \\
&\ge \lim_{t\to\infty}\frac{1}{t}\int_0^t 1(\widehat{S}_1(u)\le M)du - \lim_{t\to\infty}\frac{1}{t}\sum_{i=2}^m\int_0^t 1(\widehat{S}_i(u)>M)du \\
&= F_e(M) - (m-1)(1-F_e(M)) \ge 1 - \frac{\delta_0}{2}\,. \quad (2.3.5)
\end{aligned}$$

Because of

$$\liminf_{t\to\infty}\frac{1}{t}\int_0^t 1(\mathbf{S}(u)\in\mathbb{B}_M)du = 1 - \limsup_{t\to\infty}\frac{1}{t}\int_0^t 1(\mathbf{S}(u)\notin\mathbb{B}_M)du\,,$$

we obtain from (2.3.5) the following upper bound:

$$\limsup_{t\to\infty}\frac{1}{t}\int_0^t 1(\mathbf{S}(u)\notin\mathbb{B}_M)\,du \le \frac{\delta_0}{2}\,. \quad (2.3.6)$$

Consider the set

$$\mathbb{B}_0 = [0,\, m-1] \times \mathbb{B}_M \in \mathbb{R}_+^{m+1}\,,$$

and denote by

$$\mathcal{M}_0(t) = \int_0^t 1(\mathbf{Z}(u)\in\mathbb{B}_0)\,du\,,$$

the amount of time that the process $\{\mathbf{Z}(t)\}$ stays in the set $\mathbb{B}_0$ during the interval $[0,\ t]$. Relying on the inequality

$$\mathsf{1}(\mathbf{Z}(u) \in \mathbb{B}_0) \geq \mathsf{1}(\nu(u) < m) - \mathsf{1}(\mathbf{S}(u) \notin \mathbb{B}_M)\,, \tag{2.3.7}$$

we obtain from (2.2.6), (2.3.6) and (2.3.7) that

$$\liminf_{t\to\infty} \frac{\mathcal{M}_0(t)}{t} \geq \liminf_{t\to\infty} \frac{I(t)}{t} - \limsup_{t\to\infty} \frac{1}{t} \int_0^t \mathsf{1}(\mathbf{S}(u) \notin \mathbb{B}_M)\,du \geq \frac{\delta_0}{2}\,. \tag{2.3.8}$$

Furthermore, evidently,

$$G_0(t) := \sum_{k=1}^{A(t)} \mathsf{1}(\mathbf{Z}_k \in \mathbb{B}_0) = \sum_{k=1}^{A(t)} \mathsf{1}(\nu_k \in [0,\ m),\ \mathbf{S}(t_k^-) \in \mathbb{B}_M)\,, \tag{2.3.9}$$

represents the number of arrivals, within the interval $[0,\ t]$, that observe the process $\{\mathbf{Z}_k\}$ in the set $\mathbb{B}_0$. It follows from (2.3.8) that $\mathcal{M}_0(t) \to \infty$ w.p.1.

Now we show that $G_0(t) \to \infty$ as well. By construction, all components of the process $\{\mathbf{Z}_k\}$ may only decrease in between arrivals, i.e., in any interval $(t_k,\ t_{k+1})$, $k \geq 1$. In particular, if $\mathbf{Z}_k \in \mathbb{B}_0$ for some k, then the time that the process spends in the set $\mathbb{B}_0$ during the interval $(t_k,\ t_{k+1})$ is upper bounded by its length τ_k (this is illustrated in Fig. 2.6). Using (2.3.9), this allows to establish the following inequality relating the time $\mathcal{M}_0(t)$ and the number of arrivals $G_0(t)$:

$$\begin{aligned}\mathcal{M}_0(t) &= \sum_{k=1}^{A(t)-1} \int_{t_k}^{t_{k+1}} \mathsf{1}(\mathbf{Z}(u) \in \mathbb{B}_0)du + \int_{t_{A(t)}}^{t} \mathsf{1}(\mathbf{Z}(u) \in \mathbb{B}_0)du \\ &\leq \sum_{k=1}^{A(t)-1} \mathsf{1}(\mathbf{Z}_{k+1} \in \mathbb{B}_0)\tau_k + \mathsf{1}(\mathbf{Z}_{A(t)+1} \in \mathbb{B}_0)\tau_{A(t)} \\ &= \sum_{k=1}^{A(t)} \mathsf{1}(\mathbf{Z}_{k+1} \in \mathbb{B}_0)\tau_k \leq (G_0(t)+1) \max_{1\leq k\leq A(t)} \tau_k,\quad t \geq 0\,.\end{aligned} \tag{2.3.10}$$

Figure 2.6 illustrates the inequality (2.3.10) for the one-dimensional component $\mathbf{S}(t) = S(t)$ of the process $\{\mathbf{Z}(t)\}$ corresponding to a single-server system, in which case $\mathbb{B}_M = [0,\ M]$.

Next, we prove that the iid interarrival times $\{\tau_k\}$ and the number of arrivals $A(t)$ satisfy

$$\lim_{t\to\infty} \frac{1}{t} \max_{k\leq A(t)} \tau_k = 0 \quad \text{w.p.1}\,. \tag{2.3.11}$$

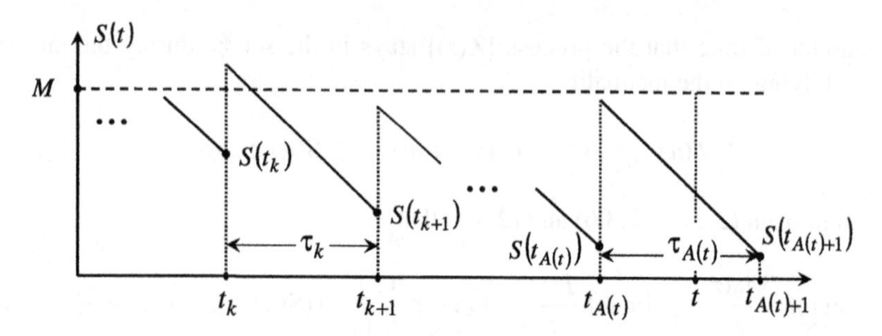

Fig. 2.6 The dynamics of the one-dimensional component $\mathbf{S}(t) = S(t)$ of the process $\{\mathbf{Z}(t)\}$

First of all we note that (2.3.11) can also be written as

$$\lim_{t\to\infty} \frac{1}{t} \max_{k\le A(t)} \mathcal{T}_k = \lim_{t\to\infty} \left[\frac{A(t)}{t} \max_{k\le A(t)} \frac{\mathcal{T}_k}{A(t)}\right] = \lambda \lim_{t\to\infty} \max_{k\le A(t)} \frac{\mathcal{T}_k}{A(t)} = 0\,,$$

and therefore (2.3.11) is equivalent to

$$\lim_{n\to\infty} \frac{1}{n} \max_{k\le n} \mathcal{T}_k = 0 \quad \text{w.p.1}\,. \tag{2.3.12}$$

Since $\mathcal{T}_n = t_{n+1} - t_n$ and $\mathsf{E}\mathcal{T} < \infty$, then by invoking the SLLN, w.p.1

$$\lim_{n\to\infty} \frac{\mathcal{T}_n}{n} = \lim_{n\to\infty} \left[\frac{t_{n+1}}{n+1}\frac{n+1}{n} - \frac{t_n}{n}\right] = \mathsf{E}\mathcal{T} - \mathsf{E}\mathcal{T} = 0\,.$$

It is known that (see [4], Chap. 2, Sect. 10)

$$\lim_{n\to\infty} \frac{\mathcal{T}_n}{n} = 0 \;\; \text{w.p.1}\,,$$

if and only if for arbitrary constants $\varepsilon,\ \delta > 0$, there exists an integer n_0 such that

$$\mathsf{P}(\max_{k\ge n} \frac{\mathcal{T}_k}{k} > \varepsilon) \le \delta\,, \quad n \ge n_0\,. \tag{2.3.13}$$

On the other hand, for $n \ge n_0$,

$$\begin{aligned}
\mathsf{P}\Big(\frac{1}{n} \max_{1\le k\le n} \mathcal{T}_k > \varepsilon\Big) &\le \mathsf{P}\Big(\frac{t_{n_0}}{n} + \max_{n_0\le k\le n} \frac{\mathcal{T}_k}{k} > \varepsilon\Big) \\
&\le \mathsf{P}\Big(\frac{t_{n_0}}{n} + \max_{n_0\le k\le n} \frac{\mathcal{T}_k}{k} > \varepsilon,\ \frac{t_{n_0}}{n} \le \frac{\varepsilon}{2}\Big) + \mathsf{P}\Big(\frac{t_{n_0}}{n} > \frac{\varepsilon}{2}\Big) \\
&\le \mathsf{P}\Big(\max_{k\ge n_0} \frac{\mathcal{T}_k}{k} > \frac{\varepsilon}{2}\Big) + \mathsf{P}\Big(\frac{t_{n_0}}{n} > \frac{\varepsilon}{2}\Big)\,.
\end{aligned} \tag{2.3.14}$$

Because $t_{n_0} < \infty$ w.p.1, then by taking n large enough we can make the 2*nd* term in (2.3.14) arbitrarily small. Since ε, δ were arbitrarily chosen, then (2.3.12) follows from (2.3.13), (2.3.14). Consequently, it follows from (2.3.8)–(2.3.11) that

$$G_0(t) \to \infty \text{ w.p.1}. \tag{2.3.15}$$

To conclude this part of the proof, we must return from the process $\{\mathbf{Z}_k\}$ to the original process $\{\mathbf{Y}_k\}$. As indicated before, $\tau(t_k^-) = 0,\ k \geq 1$. Also note that if $\mathbf{S}(t) \in \mathbb{B}_M$ and $\nu(t) < m$, then the inequality $\mathsf{W}(t) \leq mM$ holds. Hence, if $\mathbf{Z}_k \in \mathbb{B}_0$, then

$$\mathbf{Y}_k \in [0,\ m-1] \times [0,\ mM] \times \{0\} =: \mathbb{B}.$$

In other words, $\{\mathbf{Z}_k \in \mathbb{B}_0\} \subseteq \{\mathbf{Y}_k \in \mathbb{B}\}$, and, relying on (2.3.15), we obtain

$$\sum_{k=1}^{A(t)} \mathsf{1}(\mathbf{Y}_k \in \mathbb{B}) \to \infty, \tag{2.3.16}$$

that is, the total number of visits by the process $\{\mathbf{Y}_k\}$ to the set $\mathbb{B}$ grows unlimitedly w.p.1 as time increases, for any initial state $\mathbf{Y}_1$.

Step 2. Next, we prove that the *mean number* of visits to the set $\mathbb{B}$ by the process $\{\mathbf{Y}_k\}$ within the 1*st* regeneration cycle is *finite*. Let $L = \lceil mM/\varepsilon_0 \rceil$, where ε_0 satisfies (2.2.11). Then from (2.2.19), we obtain for each integer $k \geq 1$ that

$$\mathsf{P}(\theta_1 \leq (k+1)L \mid \mathbf{Y}_{kL} \in \mathbb{B},\ \theta_1 > kL) = \mathsf{P}(\theta(kL) \leq L \mid \mathbf{Y}_{kL} \in \mathbb{B},\ \theta_1 > kL) \geq \delta_1^L.$$

Therefore, we can derive the following chain of the inequalities

$$\begin{aligned} 1 &\geq \sum_{k=1}^{\infty} \mathsf{P}(kL < \theta_1 \leq (k+1)L,\ \mathbf{Y}_{kL} \in \mathbb{B}) \\ &= \sum_{k=1}^{\infty} \mathsf{P}(\theta(kL) \leq L \mid \mathbf{Y}_{kL} \in \mathbb{B},\ \theta_1 > kL)\, \mathsf{P}(\theta_1 > kL,\ \mathbf{Y}_{kL} \in \mathbb{B}) \\ &\geq \delta_1^L \sum_{k=1}^{\infty} \mathsf{P}(\theta_1 > kL,\ \mathbf{Y}_{kL} \in \mathbb{B}). \end{aligned} \tag{2.3.17}$$

If $L = 1$, then

$$\sum_{k=1}^{\infty} \mathsf{P}(\theta_1 > kL,\ \mathbf{Y}_{kL} \in \mathbb{B}) \equiv \sum_{k=1}^{\infty} \mathsf{P}(\theta_1 > k,\ \mathbf{Y}_k \in \mathbb{B}) = \mathsf{E} \sum_{k=1}^{\theta_1 - 1} \mathsf{1}(\mathbf{Y}_k \in \mathbb{B}) =: G(\mathbb{B}),$$

is the mean number of arrivals, within the $1st$ regeneration cycle, observing the process $\{\mathbf{Y}_k\}$ in the set $\mathbb{B}$, and it follows from (2.3.17) that

$$G(\mathbb{B}) \le \frac{1}{\delta_1} .$$

Otherwise, if $L > 1$, then continuing as in (2.3.17), we obtain for each $\ell = 0, \dots, L-1$, that

$$1 \ge \delta_1^L \sum_{k=1}^{\infty} \mathsf{P}(\mathbf{Y}_{k\,L+\ell} \in \mathbb{B},\ \theta_1 > k\,L + \ell) . \tag{2.3.18}$$

Now, summing up the inequalities in (2.3.18) over $\ell = 0, \dots, L-1$, yields the bound

$$\begin{aligned}
L &\ge \delta_1^L \sum_{\ell=0}^{L-1} \sum_{k=1}^{\infty} \mathsf{P}(\mathbf{Y}_{k\,L+\ell} \in \mathbb{B},\ \theta_1 > k\,L + \ell) \\
&= \delta_1^L \sum_{k=1}^{\infty} \sum_{\ell=0}^{L-1} \mathsf{P}(\mathbf{Y}_{k\,L+\ell} \in \mathbb{B},\ \theta_1 > k\,L + \ell) \\
&= \delta_1^L \sum_{k=L}^{\infty} \mathsf{P}(\mathbf{Y}_k \in \mathbb{B},\ \theta_1 > k) = \delta_1^L\ \mathsf{E} \sum_{k=L}^{\theta_1 - 1} \mathsf{1}(\mathbf{Y}_k \in \mathbb{B}) .
\end{aligned} \tag{2.3.19}$$

Consequently, we find

$$\mathsf{E} \sum_{k=L}^{\theta_1 - 1} \mathsf{1}(\mathbf{Y}_k \in \mathbb{B}) \le \frac{L}{\delta_1^L} . \tag{2.3.20}$$

Taking into account the $L-1$ values of $\mathbf{Y}_k$ for $k = 1, \dots, L-1$, we obtain from (2.3.19) and (2.3.20) that

$$G(\mathbb{B}) \equiv \mathsf{E} \sum_{k=1}^{\theta_1 - 1} \mathsf{1}(\mathbf{Y}_k \in \mathbb{B}) \le \mathsf{E} \sum_{k=L}^{\theta_1 - 1} \mathsf{1}(\mathbf{Y}_k \in \mathbb{B}) + L - 1 < L\Big(1 + \frac{1}{\delta_1^L}\Big) . \tag{2.3.21}$$

Therefore, the *mean number of arrivals* that observe the process $\{\mathbf{Y}_k\}$ in the set $\mathbb{B}$ *within the* $1st$ *regeneration cycle* is finite, and as a result, the number of such visits satisfies

$$\sum_{k=1}^{\theta_1 - 1} \mathsf{1}(\mathbf{Y}_k \in \mathbb{B}) < \infty \ \text{w.p.1} ,$$

for each initial state $\mathbf{Y}_1$. Combined with (2.3.16), this implies that the total number of regeneration cycles is *no less than two*, implying $\theta_1 < \infty$, and hence $T_1 < \infty$, w.p.1. □

Therefore, the conditions $\rho < m$ and $\mathsf{P}(\tau > S) > 0$ both form the *stability criteria* of the $GI/G/m$ system (in terms of *classical regenerations*). It also follows from the above analysis that the number of regeneration cycles is indeed infinite w.p.1, or equivalently, $\theta_n < \infty$ for each $n \geq 1$, also see [5].

Problem 2.3 *Explain the* $1st$ *inequality in (2.3.5) using the inequality*

$$\mathsf{P}(\mathbf{S}(t) \in \mathbb{B}_M) \equiv \mathsf{P}(S_1(t) \leq M,\ S_2(t) \leq M) \leq \mathsf{P}(S_1(t) \leq M) - \mathsf{P}(S_2(t) > M)\,,$$

for the two-dimensional process $\mathbf{S}(t) = (S_1(t),\ S_2(t)),\ t \geq 0$.

Now we prove the following statement which is similar to result (1.2.17) and repeatedly used in our analysis.

Theorem 2.4 *In the positive recurrent system, w.p.1,*

$$\lim_{t\to\infty} \frac{\mathsf{W}(t)}{t} = 0\,, \tag{2.3.22}$$

$$\lim_{t\to\infty} \frac{\nu(t)}{t} = 0\,. \tag{2.3.23}$$

Proof Since the remaining regeneration time satisfies $T(t) \geq W_i(t)$ for each i, then also $mT(t) \geq \mathsf{W}(t)$, implying (2.3.22) due to (1.2.17). Note that the queue-size process $\{\nu(t)\}$ can be treated as a process with *zero* regenerative increments, $\nu(T_{n+1}) - \nu(T_n) \equiv 0$, and the maximum over a regeneration cycle is (stochastically) upper bounded by the discrete-time cycle length θ,

$$\max_{0\leq t<T} \nu(t) \leq_{st} \theta\,.$$

Then (2.3.23) follows from (1.2.16) because $\mathsf{E}\theta < \infty$. Note that the same argument can be applied to the process $\{\mathsf{W}(t)\}$ as well, which has *zero mean* increments,

$$\mathsf{E}[\mathsf{W}(T_{n+1}) - \mathsf{W}(T_n)] = \mathsf{E}[S_{\theta_{n+1}} - S_{\theta_n}] \equiv 0\,,$$

and the mean cycle maximum which is upper bounded by $\mathsf{E}S\ \mathsf{E}\theta$. □

2.4 Instability

In this section we study in more detail the instability conditions of the $GI/G/m$ system considered in Sects. 2.2 and 2.3. Denote by $Q(t)$ the *queue size* at instant t.

Theorem 2.5 *(i) If $\rho > m$, then $\mathsf{E}T = \infty$ and $\nu(t)$, $\mathsf{W}(t) \to \infty$ w.p.1, irrespective of the initial state. (ii) If $\rho = m$ and $\mathsf{W}_1 = 0$, then $\mathsf{E}T = \infty$.*

Proof (*i*) Denote by $\widehat{D}_i(t)$ the number of renewals in the interval $[0,\ t]$ in the zero-delayed renewal process generated by successive service times in server i, and consider the aggregated process

$$\widehat{D}(t) = \sum_{i=1}^{m} \widehat{D}_i(t),\ t \geq 0\,.$$

Observe that $A(t)$ and $D(t)$, the total number of arrivals and departures in $[0,\ t]$ respectively, satisfy the relations

$$\nu(t) = \nu_1 + A(t) - D(t) \geq A(t) - \widehat{D}(t)\ ,\ t \geq 0\,. \tag{2.4.1}$$

Because, by the SLLN for renewal processes (1.2.10),

$$\lim_{t\to\infty} \frac{\widehat{D}(t)}{t} = m\mu,\quad \lim_{t\to\infty} \frac{A(t)}{t} = \lambda\,, \tag{2.4.2}$$

then we immediately obtain from (2.4.1)

$$\liminf_{t\to\infty} \frac{\nu(t)}{t} \geq \mu(\rho - m) > 0\,, \tag{2.4.3}$$

also implying $\mathsf{E}T = \infty$. Note that

$$Q(t) = (\nu(t) - m)^+ \geq \nu(t) - m\ ,$$

due to (2.4.3) implying

$$\liminf_{t\to\infty} \frac{Q(t)}{t} \geq \mu(\rho - m)\ , \tag{2.4.4}$$

and, in particular, $Q(t) \to \infty$ w.p.1. Denote by $\mathcal{G}(t)$ the set of customers waiting in the *queue* (i.e., not including the servers) at time t, so the cardinality of this set $|\mathcal{G}(t)| = Q(t)$. Then, since $\mathsf{W}(t) \geq \sum_{k\in\mathcal{G}(t)} S_k$, we also obtain, by (2.4.4) and the SLLN,

$$\liminf_{t\to\infty} \frac{\mathsf{W}(t)}{t} \geq \liminf_{t\to\infty} \frac{\sum_{k\in\mathcal{G}(t)} S_k}{Q(t)} \frac{Q(t)}{t} \geq \rho - m\,. \tag{2.4.5}$$

(ii) In this (zero-delayed) case assume that $\mathsf{E}T < \infty$. Then assumption $\mathsf{P}(\tau > S) > 0$ holds, and it follows as in (2.1.18)–(2.1.20) from balance equation

$$V(t) = \mathsf{W}(t) + mt - \widehat{I}(t)\,, \tag{2.4.6}$$

that

$$\lim_{t\to\infty} \frac{V(t)}{t} \equiv \rho = m - \frac{\mathsf{E}\widehat{I}_0}{\mathsf{E}T} < m\,, \tag{2.4.7}$$

because $\mathsf{W}(t) = o(t)$ in view of (2.3.22). Consequently, the contradiction with assumption $\rho = m$ implies $\mathsf{E}T = \infty$. □

The results in (2.4.3)–(2.4.5) describe the so-called *transient* regenerative processes, in which case the renewal process of regenerations $\{T_n\}$ is transient or terminating [6, 7].

Problem 2.4 *Let $\nu(t) \Rightarrow \infty$. Prove that $\rho \geq m$. Hint: show that $\mathsf{E}\widehat{I}(t) = o(t)$ and use relation (2.4.6).*

The following statement complements Theorem 2.5. It shows that in the *boundary case* $\rho = m$ the basic processes diverge to infinity in probability.

Theorem 2.6 *(i) If $\rho = m$, $\mathsf{W}_1 = 0$ and $P(\tau > S) > 0$, then as $t \to \infty$,*

$$\nu(t) \Rightarrow \infty\,,\ \ \mathsf{W}(t) \Rightarrow \infty\,,$$
$$\mathsf{E}\nu(t) = o(t)\,,\ \ \mathsf{EW}(t) = o(t)\,.$$

(ii) If $\nu(t) \Rightarrow \infty$, then $\mathsf{W}(t) \Rightarrow \infty$ and $\rho \geq m$.
(iii) If $\mathsf{W}(t) \Rightarrow \infty$, then $\nu(t) \Rightarrow \infty$ and $\rho \geq m$.

Proof (i) Assume that $\nu(t) \not\Rightarrow \infty$, implying

$$\inf_{n\geq 1} \mathsf{P}(\nu(z_n) \leq C) > 0\,, \tag{2.4.8}$$

for some constant C and a deterministic time sequence $z_n \to \infty$. Conditioned on the event $\{\nu(z_n) \leq C\}$, we obtain an evident (stochastic) upper bound

$$\mathsf{W}(z_n) \leq_{st} \sum_{k=1}^{C} S_k + \sum_{i=1}^{m} S_i(z_n)\,, \tag{2.4.9}$$

where the aggregated remaining service time process $\{\sum_{i=1}^{m} S_i(z_n),\ n \geq 1\}$ is tight, in view of Theorem 3.3 in Sect. 3.1. In the above (stochastic) inequality, we take into account that, among at most C customers present in the system at instant z_n, there are at most m customers that may be in service. It then follows from (2.4.8) that there exists a constant D such that the following lower bound (uniform in n) holds:

$$\inf_n \mathsf{P}(\mathsf{W}(z_n) \le D) \ge \inf_n \mathsf{P}\Big(\nu(z_n) \le C, \sum_{k=1}^{C} S_k \le D/2, \sum_{i=1}^{m} S_i(z_n) \le D/2\Big) > 0 .$$

Then $\mathsf{E}T < \infty$ follows as in Theorem 2.2. In this case $\mathsf{W}(t) = o(t)$, and we obtain $\rho < m$ by invoking (2.4.7). This contradiction shows that $\nu(t) \Rightarrow \infty$. In addition, for each $x \ge 0$ and integer $N \ge 1$, we find that

$$\begin{aligned} \mathsf{P}(\mathsf{W}(t) > x) &\ge \mathsf{P}\Big(\sum_{k=1}^{N} S_k > x,\ Q(t) = \nu(t) - m \ge N\Big) \\ &\ge \mathsf{P}\Big(\sum_{k=1}^{N} S_k > x\Big) - \mathsf{P}(\nu(t) < N + m) . \end{aligned} \tag{2.4.10}$$

For arbitrary (fixed) x and $\varepsilon > 0$, one can choose N such that

$$\mathsf{P}(\sum_{k=1}^{N} S_k > x) \ge 1 - \varepsilon/2 .$$

Then, in view of $\nu(t) \Rightarrow \infty$, we can choose t_0 such that

$$\mathsf{P}(\nu(t) < N + m) \le \varepsilon/2,\ t \ge t_0 .$$

This, in turn, in combination with (2.4.10), implies that

$$\mathsf{P}(\mathsf{W}(t) > x) \ge 1 - \varepsilon, \quad t \ge t_0 ,$$

and since x and ε were arbitrarily chosen, then $\mathsf{W}(t) \Rightarrow \infty$ immediately follows. As a result,

$$\mathsf{P}(S_i(t) = 0) \to 0\ ,\ t \to \infty\ ,\ i = 1, \ldots, m\ ,$$

and the mean total idle time satisfies

$$\lim_{t\to\infty} \frac{1}{t}\, \mathsf{E}\widehat{I}(t) = \frac{1}{t} \int_0^t \sum_{i=1}^{m} \mathsf{P}(S_i(u) = 0)du = 0 . \tag{2.4.11}$$

In addition, Eq. (2.4.6), written as

$$\mathsf{E}\mathsf{W}(t) = \mathsf{E}V(t) - mt - \mathsf{E}\widehat{I}(t) , \tag{2.4.12}$$

and (2.4.7), (2.4.11) lead to

$$\lim_{t\to\infty} \frac{1}{t}\ \mathsf{E}\mathsf{W}(t) = \rho - m = 0\,. \tag{2.4.13}$$

In order to prove $\mathsf{E}\nu(t) = o(t)$, we note that,

$$\mathsf{W}(t) \geq_{st} \sum_{k=1}^{Q(t)} S_k = \sum_{k=1}^{\infty} S_k\ \mathsf{1}(Q(t) \geq k)\,,$$

where the stochastic inequality is used for the same reason as in (2.4.9). Since S_k and $\mathsf{1}(Q(t) \geq k)$ are independent, then

$$\mathsf{E}\mathsf{W}(t) \geq \sum_{k\geq 1} \mathsf{E}\big[S_k\ \mathsf{1}(Q(t) \geq k)\big] = \mathsf{E}S\ \mathsf{E}Q(t)\,,$$

implying, due to (2.4.13)

$$\lim_{t\to\infty} \frac{\mathsf{E}Q(t)}{t} = 0\,, \quad \lim_{t\to\infty} \frac{\mathsf{E}\nu(t)}{t} = 0\,.$$

(ii) Above we have shown that $\nu(t) \Rightarrow \infty$ implies $\mathsf{W}(t) \Rightarrow \infty$, and $\rho \geq m$ follows from (2.4.12).
(iii) Assume that $\mathsf{W}(t) \Rightarrow \infty$ but $\nu(t) \not\Rightarrow \infty$. Then, similar as in the discussion following formula (2.4.8), we arrive at $\mathsf{E}T < \infty$, further leading to $\rho < m$, see (2.4.7). This contradiction shows that indeed $\nu(t) \Rightarrow \infty$, implying (2.4.11). The inequality $\rho \geq m$ then follows from (2.4.12), also see Problem 2.4. □

2.5 Stationary Performance Measures

The main limit results for regenerative processes, represented by (1.2.5), (1.2.14), (1.2.15) and (2.1.7), do not only demonstrate the critical importance of the finiteness of the mean cycle length for the stability analysis of a system, but as we show below, also are a powerful tool in obtaining some important stationary performance measures, often easily and in an explicit form.

2.5.1 Little's Laws

Consider the $GI/G/m$ system above under assumptions $\rho < m$ and $\mathsf{P}(\tau > S) > 0$ implying positive recurrence. We now consider the *sojourn time* V_n of the n-th customer which is defined as

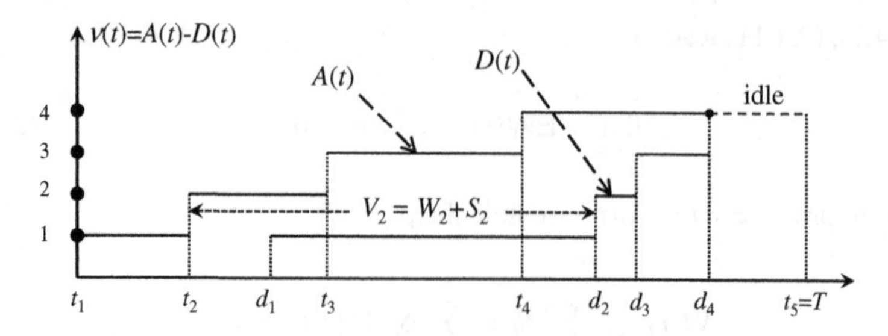

Fig. 2.7 Illustration of equality (2.5.2) leading to Little's law (2.5.4)

$$V_n = W_n + S_n, \ n \geq 1 ,$$

where, as in (2.1.1), $W_n = W(t_n^-)$ is the waiting time of customer n in the queue. Since the regeneration period length θ is aperiodic, then the weak limit $W_n \Rightarrow W$ exists, and the limit

$$V_n \Rightarrow V =_{st} W + S ,$$

exists as well and represents the stationary sojourn time of a customer. Since the sequence $\{W_n + S_n\}$ is regenerative, then we can deduce from (2.1.7) that

$$\frac{\mathsf{E}\sum_{n=1}^{\theta}(W_n + S_n)}{\mathsf{E}\theta} = \mathsf{E}V = \mathsf{E}W + \mathsf{E}S . \tag{2.5.1}$$

On the other hand, it follows from the geometrical properties of the sample paths of the involved processes that

$$\int_0^T \nu(t)dt = \int_0^T \big(A(t) - D(t)\big)dt = \sum_{n=1}^{\theta}(W_n + S_n) , \tag{2.5.2}$$

which is illustrated by Fig. 2.7, and that

$$\frac{\mathsf{E}\int_0^T \nu(t)dt}{\mathsf{E}T} = \lambda \frac{\mathsf{E}\sum_{n=1}^{\theta}(W_n + S_n)}{\mathsf{E}\theta} , \tag{2.5.3}$$

where Wald's identity $\mathsf{E}T = \mathsf{E}\theta \ \mathsf{E}\tau = \mathsf{E}\theta/\lambda$ was also applied. If regeneration period T is non-lattice (and in particular, if the input process is Poisson), then the weak limit $\nu(t) \Rightarrow \nu$ exists as well. Then, from expressions (2.5.3) and (2.5.1), a well-known representation of *Little's law*, namely

$$\mathsf{E}\nu = \lambda \, \mathsf{E}V , \tag{2.5.4}$$

follows, relating the mean stationary sojourn time at *arrival instants* and the mean stationary number of customers in the system at *arbitrary instants* (that is in 'continuous' time).

Remark 2.4 According to (1.2.15), relation (2.5.4) and analogous relations below allows both sides to be in infinite. Indeed, the finiteness of the these stationary measures requires an extra moment assumption $\mathsf{E}S^2 < \infty$, see the next Sect. 2.5.2.

Using the same arguments, it is then easy to establish the following expression for Little's law relating the mean stationary *queue size* $\mathsf{E}Q$ (the number of waiting customers in the system, excluding the servers) in continuous time and the mean stationary remaining work $\mathsf{E}W$ observed by an arriving customer

$$\mathsf{E}Q = \lambda\,\mathsf{E}W\,. \tag{2.5.5}$$

Therefore, we may summarize (2.5.1)–(2.5.5) as follows:

Theorem 2.7 *Under the preceding conditions, the following variants of the Little's law hold:*

$$\begin{aligned} \mathsf{E}\nu &= \lambda \mathsf{E}V = \lambda \mathsf{E}W + \rho\,, \\ \mathsf{E}Q &= \lambda \mathsf{E}W\,. \end{aligned} \tag{2.5.6}$$

Problem 2.5 *Prove equality (2.5.5) using (2.5.1).*

Let $A_s(t)$ be the number of customers entering the servers in the interval $[0,\ t]$, then $\Theta(t) := A_s(t) - D(t)$ is the number of customers being served at instant t. Define the weak limit $\Theta(t) \Rightarrow \Theta$.

Problem 2.6 *Using a properly modified version of expressions (2.5.2) and (2.5.3), prove the following variant of Little's law:*

$$\mathsf{E}\Theta = \lambda \mathsf{E}S = \rho\,,$$

which is in agreement with (2.5.6) because of $\mathsf{E}\nu = \mathsf{E}Q + \mathsf{E}\Theta$.

2.5.2 *PASTA and the Pollaczek-Khintchine Formula*

Another important and insightful case study, presented in this subsection, concerns the $M/G/1$ queueing system in which the positive recurrence condition $\rho = \lambda\,\mathsf{E}S < 1$ is satisfied, and the additional (moment) assumption $\mathsf{E}S^2 < \infty$ holds as well. The regeneration period length T is non-lattice, and then there exists the *stationary* workload process

$$W(t) \Rightarrow \widehat{W}, \ t \to \infty . \tag{2.5.7}$$

Again, relying on a geometrical consideration of the sample path, we obtain from (1.2.15) and (2.5.7) that

$$\begin{aligned} \mathsf{E}\widehat{W} = \frac{1}{\mathsf{E}T}\,\mathsf{E}\int_0^T W(t)dt &= \lambda \frac{\mathsf{E}\big[\sum_{n=1}^{\theta}\big((W_n+S_n)^2/2 - W_{n+1}^2/2\big)\big]}{\mathsf{E}\,\theta} \\ &= \lambda \frac{\mathsf{E}\big[\sum_{n=1}^{\theta}(W_nS_n + S_n^2/2)\big]}{\mathsf{E}\,\theta} . \end{aligned} \tag{2.5.8}$$

In the above derivation, Fig. 2.8 explains the equality

$$\int_0^T W(t)dt = \sum_{n=1}^{\theta}\Big(\frac{1}{2}(W_n+S_n)^2 - \frac{1}{2}W_{n+1}^2\Big),$$

which equals the surface under the curve of $W(t)$, calculated over a regeneration cycle. (Note that in the example presented in Fig. 2.8, $\theta = 3$, and $W_1 = W_4 = 0$.)

Note that the sequence $\{W_nS_n + S_n^2/2,\ n \geq 1\}$, classically regenerates, and is distributed as $S^2/2$ at the instants θ_k because $W_{\theta_k} = 0$. This is an example of a classical regeneration with a non-degenerated distribution, although in most of cases that we consider this distribution is concentrated at zero (see also the related discussion in Sect. 2.6). Condition $\rho < 1$ and the aperiodicity of θ imply that the following weak limit exists

$$W_nS_n + \frac{S_n^2}{2} \Rightarrow WS + \frac{S^2}{2}, \ n \to \infty , \tag{2.5.9}$$

and we find from (2.1.7) that

$$\mathsf{E}\Big[WS + \frac{S^2}{2}\Big] = \frac{\mathsf{E}\sum_{n=1}^{\theta}(W_nS_n + S_n^2/2)}{\mathsf{E}\,\theta} . \tag{2.5.10}$$

Now, observe that (2.5.8) and (2.5.10) imply the relation

$$\mathsf{E}\widehat{W} = \lambda\,\mathsf{E}\Big[SW + \frac{S^2}{2}\Big] = \rho\ \mathsf{E}W + \lambda\frac{\mathsf{E}S^2}{2} , \tag{2.5.11}$$

where the independence between W and S has been invoked.

Remark 2.5 In the $M/G/1$ system with $\rho < 1$ the stationary workload process $\widehat{W}$ exists but, as we can deduce from expression (2.5.11), its mean is infinite if $\mathsf{E}S^2 = \infty$.

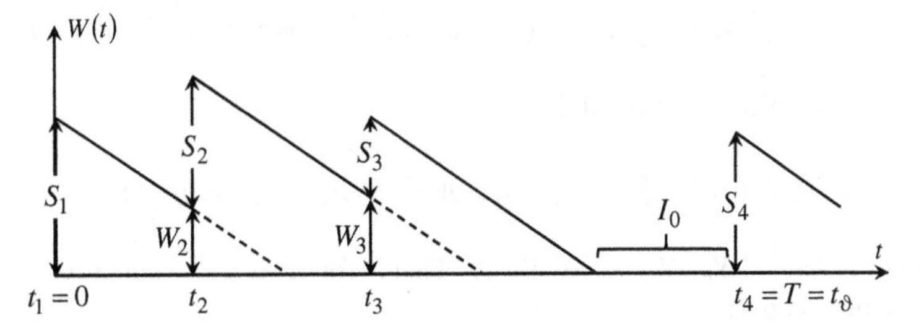

Fig. 2.8 The path of the workload process $W(t)$ over a regeneration cycle

Our next goal is to prove the following stochastic equality:

$$\widehat{W} =_{st} W \,. \tag{2.5.12}$$

This is a particular case of the *PASTA* property [8] which states that (in a stationary regime) an arriving customer observes the same amount of remaining work as would be observed at an arbitrary instant t.

Remark 2.6 From a practical point-of-view, the PASTA property allows one to equate the (limiting) *fraction of customers* which observe a given state of the system with the *fraction of time* that the system spends in this state. This is for instance especially important for multiclass (and related) systems with independent Poisson input processes, and implies that the state of the system observed by an arriving customer is *independent of the customer class*. This property is widely used below, in particular, in the analysis of retrial systems in Chaps. 7 and 8.

Note that, for the embedded Markov chain $\{W_n\}$, the following *Lindley recursion* holds:

$$W_{n+1} = (W_n + S_n - \tau_n)^+ , \ n \geq 1 \,, \tag{2.5.13}$$

which is illustrated by Fig. 2.8 as well. This recursion implies the convergence

$$W_n \Rightarrow W =_{st} (W + S - \tau)^+ \ , \ n \to \infty \,, \tag{2.5.14}$$

where we have applied the *continuous mapping theorem* [1]. Define the *attained* interarrival time at time instant $t \geq 0$ as

$$\widehat{\tau}(t) = \max_{k\geq 1}(t - t_k : t \geq t_k) \,. \tag{2.5.15}$$

As before, $A(t)$ represents the number of (Poisson) arrivals that occur in the interval $[0,\ t]$. Now we obtain the following continuous-time analogue of the Lindley recursion:

$$W(t) = \left(W_{A(t)} + S_{A(t)} - \widehat{\tau}(t)\right)^{+}, \quad t \geq 0\,, \tag{2.5.16}$$

where $W_{A(t)} := W(t_{A(t)}^{-})$. Note that

$$\widehat{\tau}(t_{n+1}) = 0, \;\; \widehat{\tau}(t_{n+1}^{-}) = \tau_n, \;\; A(t_n) = n, \;\; A(t_n^{-}) = n - 1\,,$$

implying $W_{A(t_n)} = W_n$, $W_{A(t_n^{-})} = W_{n-1}$. Then, for $t = t_{n+1}$, equality (2.5.16) becomes the identity:

$$W(t_{n+1}) \equiv W_n + S_n = (W_n + S_n)^{+},$$

while, if we take $t = t_{n+1}^{-}$, then (2.5.16) transforms into the classical Lindley recursion (2.5.13).

For our further analysis it is important to show that the (weak) limits of $W(t)$, W_n, and $W_{A(t)}$ are the same, implying the PASTA property (2.5.12). Now, for each $x \geq 0$, we can find distribution of $W_{A(t)}$ as $t \to \infty$ as follows:

$$\begin{aligned}
\lim_{t\to\infty} \mathsf{P}(W_{A(t)} \leq x) &= \lim_{t\to\infty} \frac{1}{t}\int_0^t \mathsf{1}(W_{A(u)} \leq x)du = \frac{\mathsf{E}\int_0^T \mathsf{1}(W_{A(u)} \leq x)du}{\mathsf{E}T} \\
&= \frac{\mathsf{E}\sum_{k=1}^{\theta}\int_{t_k}^{t_{k+1}} \mathsf{1}(W_{A(u)} \leq x)du}{\mathsf{E}T} = \lambda\frac{\mathsf{E}\sum_{k=1}^{\theta}\tau_k\, \mathsf{1}(W_k \leq x)}{\mathsf{E}\theta} \\
&= \frac{\mathsf{E}\sum_{k=1}^{\theta} \mathsf{1}(W_k \leq x)}{\mathsf{E}\theta}\,,
\end{aligned} \tag{2.5.17}$$

where we have made use of Wald's identity $\mathsf{E}T = \mathsf{E}\theta\ \mathsf{E}\tau$ and the independence between τ_k and $\mathsf{1}(W_k \leq x)$ to obtain the last equality. It now remains to note that $S_{A(t)} =_{st} S$ and that the remaining and attained interarrival times have the *same limit*,

$$\tau(t) \Rightarrow \tau_e\,,\;\; \widehat{\tau}(t) \Rightarrow \tau_e\,,\;\;\; t \to \infty\,,$$

(see the analogous result (3.1.6) for the service time), and due to the *memoryless property* of the exponential τ,

$$\tau_e =_{st} \tau\,. \tag{2.5.18}$$

On the other hand,

$$\lim_{n\to\infty}\frac{1}{n}\sum_{k=1}^{n} \mathsf{1}(W_k \leq x) = \frac{\mathsf{E}\sum_{k=1}^{\theta} \mathsf{1}(W_k \leq x)}{\mathsf{E}\theta} = \mathsf{P}(W \leq x)\,,$$

and we find from (2.5.17) that

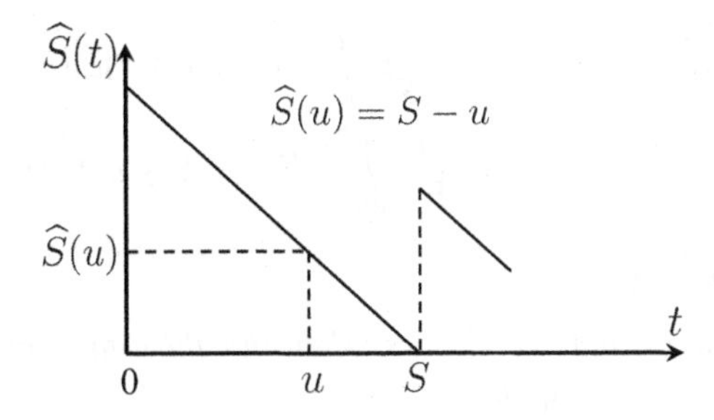

Fig. 2.9 The remaining renewal time $\widehat{S}(t)$

$$W_{A(t)} \Rightarrow W, \; t \to \infty \,. \tag{2.5.19}$$

Now PASTA (2.5.12) follows from (2.5.13), (2.5.14) and (2.5.19), because, as $t \to \infty$,

$$W(t) = \left(W_{A(t)} + S_{A(t)} - \widehat{\tau}(t)\right)^+ \Rightarrow \widehat{W} =_{st} W \,.$$

If we use this result in (2.5.11), then we finally obtain the *Pollaczek-Khintchine formula:*:

$$\mathsf{E}W = \frac{\lambda \, \mathsf{E}S^2}{2(1-\rho)} \,. \tag{2.5.20}$$

Also, let M be a random variable with geometric distribution,

$$\mathsf{P}(M = k) = (1-\rho)\rho^k, \; k \geq 0,$$

and let $\{S_e^{(k)}\}$ be independent copies of the stationary remaining renewal time S_e, with distribution function F_e defined in (2.3.3) and independent of M. The Pollaczek-Khintchine formula often appears in the more general (*distributional*) form

$$W =_{st} \sum_{k=1}^{M} S_e^{(k)} \,, \tag{2.5.21}$$

which is explained in detail, for instance, in [9]. The problem statements that follow show that expression (2.5.20) can be deduced from (2.5.21).

Problem 2.7 *Consider the remaining renewal time $\widehat{S}(t)$ in the process generated by the service times with generic element S, and let $\widehat{S}(t) \Rightarrow S_e$ be the stationary remaining renewal time. Prove that the distribution function $F_e(x) = \mathsf{P}(S_e \leq x)$ satisfies (2.3.3), using Fig. 2.9 and relations*

$$\lim_{t\to\infty}\frac{1}{t}\int_0^t 1(\widehat{S}(u)\le x)du = \frac{1}{\mathsf{E}S}\mathsf{E}\int_0^S 1(\widehat{S}(u)\le x)du$$
$$= \frac{1}{\mathsf{E}S}\int_0^\infty \mathsf{P}(u < S \le u+x)du,\ x\ge 0\,.$$

Explain the 2nd equality.

Problem 2.8 *Assuming that* $\mathsf{E}S^{n+1} < \infty$, *show that the n-th moment of the stationary remaining renewal time equals*

$$\mathsf{E}S_e^n \equiv \int_0^\infty x^n dF_e(x) = \frac{\mathsf{E}S^{n+1}}{(n+1)\,\mathsf{E}S},\quad n>0\,. \tag{2.5.22}$$

Problem 2.9 *(See [10].) Consider a stationary $M/G/1$ vacation system with* $\mathsf{E}S^2<\infty$ *in which the server, when becoming idle, takes iid random 'vacations' (with generic duration v with distribution function F_v and $\mathsf{E}v^2<\infty$) until it finds new customers waiting to be served after returning. Show that in this system the expected stationary waiting time of a customer in the queue satisfies the following modified Pollaczek-Khintchine formula (see also (2.5.20))*

$$\mathsf{E}W = \lambda\frac{\mathsf{E}S^2}{2(1-\rho)} + \frac{\mathsf{E}v^2}{2\mathsf{E}v}\,. \tag{2.5.23}$$

In particular, explain the presence of the term $\mathsf{E}v^2/(2\mathsf{E}v)$ in (2.5.23), and show that it equals the mean stationary remaining renewal (vacation) time v_e in the renewal process generated by vacations, with distribution function (see also (2.3.3))

$$\mathsf{P}(v_e\le x) = \frac{1}{\mathsf{E}v}\int_0^x (1-F_v(u))du\,,\quad x\ge 0\,.$$

Problem 2.10 *Using (2.5.22), deduce (2.5.20) from (2.5.21).*

Another interesting result for the $GI/G/1$ system with $\rho<1$ and non-lattice τ is the stationary idle probability

$$\mathsf{P}_0 = \lim_{t\to\infty}\frac{1}{t}\int_0^t 1(\nu(u)=0)du = \frac{\mathsf{E}I_0}{\mathsf{E}T} = \mathsf{P}(\nu=0) = 1-\rho\,, \tag{2.5.24}$$

where I_0 is the idle period and $\nu(t)\Rightarrow\nu$.

Problem 2.11 *Denote by B the busy period and prove the last equality in (2.5.24), by using Wald's identity and representation (see also (4.5.15))*

$$I_0 = T - B =_{st} \sum_{n=1}^{\theta}(\tau_n - S_n)\,.$$

To derive some additional results, we need the following definition. A distribution function F (of a non-negative random variable) is called *New-Better-than-Used (NBU)* if, for any $x,\ y \geq 0$, the *tail distribution* $\bar{F} = 1 - F$ satisfies the inequality

$$\bar{F}(x+y) \leqslant \bar{F}(y)\bar{F}(x) ,$$

while the opposite inequality holds for a *New-Worse-than-Used (NWU)* distribution function [11]. For instance, the *Weibull* distribution with tail

$$\bar{F}(x) = e^{-bx^{\alpha}} , \quad b > 0 , \ x \geq 0 ,$$

is NBU if $\alpha \geq 1$, and is NWU if $0 < \alpha < 1$. Note that if τ is NBU (NWU), then, with (2.5.18)

$$\tau_e \leq_{st} \tau \ (\tau_e \geq_{st} \tau) , \tag{2.5.25}$$

and that the Weibull distribution becomes exponential if $\alpha = 1$, and thus the latter distribution is NBU and NWU simultaneously.

Problem 2.12 *Assume that interarrival time τ is NBU (NWU). Relying on (2.5.25) and (2.5.11), show that $\widehat{W} \geq_{st} W$ ($\widehat{W} \leq_{st} W$), and deduce the following Pollaczek-Khintchine inequality [12, 13]:*

$$\mathsf{E}W \leq (\geq) \frac{\lambda \mathsf{E}S^2}{2(1-\rho)} .$$

Next, using the previous notations, we consider the $M/G/m$ system (with $m \geq 2$ servers) under the condition

$$\rho = \lambda \, \mathsf{E}S < m . \tag{2.5.26}$$

Let $W_i(t)$ now denote the i-th *smallest component* of the workload process vector,

$$\mathbf{W}(t) := (W_1(t), \ldots, W_m(t)) ,$$

and denote $\mathbf{W}(t_n^-) = \mathbf{W}_n,\ n \geq 1$. The sequence $\{\mathbf{W}_n\}$ constitutes an embedded Markov chain, which, under assumption (2.5.26), is a positive recurrent regenerative process satisfying the *classical Kiefer-Wolfowitz recursion* [14]

$$\begin{aligned} \mathbf{W}_{n+1} &= R\left(W_n^{(1)} + S_n - \tau_n,\ W_n^{(2)} - \tau_n, \ldots, W_n^{(m)} - \tau_n\right)^+ \\ &= R(\mathbf{W}_n + \mathbf{e}S_n - \mathbf{1}\tau_n)^+ , \quad n \geq 1 , \end{aligned} \tag{2.5.27}$$

where the operator R rearranges the components of the vector in an ascending order, and $\mathbf{e} = (1, 0, \dots, 0)$, $\mathbf{1} = (1, \dots, 1)$ are m-dimensional row vectors. Since

$$\mathsf{P}(\theta = 1) = \mathsf{P}(\tau > S) = \int_0^\infty e^{-\lambda x}\,\mathsf{P}(S \in dx) > 0\,,$$

then the regeneration period θ is aperiodic, and hence (2.5.26), (2.5.27) imply

$$\mathbf{W}_n \Rightarrow \mathbf{W} =_{st} R(\mathbf{W} + \mathbf{e}S - \mathbf{1}\tau)^+\,, \quad n \to \infty\,.$$

On the other hand, we may write the continuous-time analogue of the Kiefer-Wolfowitz representation (see also (2.5.16)) as

$$\mathbf{W}(t) = R\big(\mathbf{W}_{A(t)} + \mathbf{e}S_{A(t)} - \mathbf{1}\widehat{\tau}(t)\big)^+, \quad t \ge 0\,. \tag{2.5.28}$$

Denoting the limit $\mathbf{W}(t) \Rightarrow \widehat{\mathbf{W}}$, then by analogy with (2.5.20) we obtain the PASTA property for the stationary workload process in the $M/G/m$ system,

$$\mathbf{W} =_{st} \widehat{\mathbf{W}}\,.$$

As above, if interarrival time τ is NBU, then we obtain $\widehat{\mathbf{W}} \ge_{st} \mathbf{W}$, while for the NWU case, the relation $\widehat{\mathbf{W}} \le_{st} \mathbf{W}$ holds.

Let us consider the balance equation that follows from (2.2.14)

$$\mathsf{W}_1 + V(t) = \mathsf{W}(t) + \sum_{i=1}^{m} B_i(t)\,, \quad t \ge 0\,,$$

where $B_i(t) = t - I_i(t)$ is the busy time of server i in the interval $[0, t]$. Then it is straightforward to derive the following explicit expression for the *stationary busy probability* of an arbitrary server:

$$\mathsf{P}_B = \lim_{t\to\infty} \frac{B_i(t)}{t} = \frac{\rho}{m}\,. \tag{2.5.29}$$

This limit does not depend on the server index i, and takes the well-known form $\mathsf{P}_B = \rho$ for the single-server system.

Remark 2.7 All queueing processes that we consider in this book have piece-wise constant/linear paths and a finite w.p.1 number of jumps in any time interval $[0, t]$. Moreover, each jump is finite w.p.1 as well. The following problem shows how to prove that the workload and queue-size processes have *bounded variation* in any finite time interval.

Problem 2.13 *Assume a zero initial state. Using the geometrical properties of the sample paths of the processes* $\{\nu(t)\}$ *and* $\{\mathsf{W}(t)\}$, *show that, for each partition of the interval* $[0,\ t]$,

$$0 = z_0 < \cdots < z_n \le t\,,\quad n \ge 1\,,$$

$$\sum_{i=0}^{n-1} |\nu(z_{i+1}) - \nu(z_i)| \le A(t) + D(t)\,,$$
$$\sum_{i=0}^{n-1} |\mathsf{W}(z_{i+1}) - \mathsf{W}(z_i)| \le 2\sum_{k=1}^{A(t)} S_k\,, \tag{2.5.30}$$

where $A(t)$ $(D(t))$ *is the number of arrivals (departures) in the interval* $[0,\ t]$. *Explain why the right-hand side of each inequality (2.5.30) represents the finite w.p.1 total variation of the corresponding process in the interval* $[0,\ t]$.

Define the attained regeneration time at instant t as

$$\widetilde{T}(t) = \max_{k\ge 0}(t - T_k : t \ge T_k),\quad t \ge 0\ \ (T_0 := 0)\,. \tag{2.5.31}$$

Then the length of the regeneration cycle covering instant t is

$$T^0(t) := T(t) + \widetilde{T}(t),\quad t \ge 0\,.$$

Note that, for all $t \in [T_{k-1},\ T_k)$, it follows that $T^0(t) = T_k - T_{k-1},\ k \ge 1$, and hence, for each $x \ge 0$,

$$\mathbb{1}(T^0(t) > x) = 1 \text{ if and only if } T_k - T_{k-1} > x\,,$$

see Fig. 2.10. Now we assume that the weak limit $T^0(t) \Rightarrow T^0$ exists, and that $\mathsf{E}T^2 < \infty$. Then it follows by a standard argument that

$$\begin{aligned}\mathsf{P}(T^0 > x) &= \lim_{t\to\infty}\frac{1}{t}\int_0^t \mathbb{1}(T^0(u) > x)du = \frac{\mathsf{E}\int_0^T \mathbb{1}(T^0(u) > x)du}{\mathsf{E}T}\\ &= \frac{\mathsf{E}[T\,\mathbb{1}(T > x)]}{\mathsf{E}T} = \frac{1}{\mathsf{E}T}\int_x^\infty u\,\mathsf{P}(T \in du)\,. \end{aligned}\tag{2.5.32}$$

Problem 2.14 *Denoting by* $F(x) = \mathsf{P}(T \le x)$, *show that the distribution (2.5.32) can be represented as*

$$\mathsf{P}(T^0 > x) = \frac{1}{\mathsf{E}T}\Big[x\bar{F}(x) + \int_x^\infty \bar{F}(u)du\Big],\ x \ge 0\,. \tag{2.5.33}$$

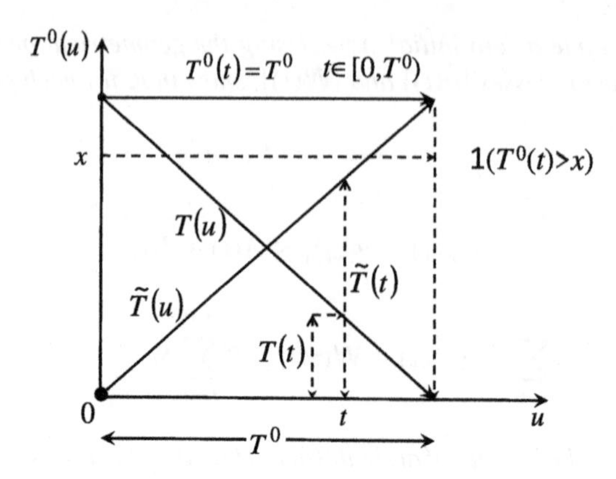

Fig. 2.10 Illustration of $T^0(t) = T(t) + \widetilde{T}(t) = T^0$ in steady state

Using (2.5.33) and representation

$$\mathsf{E}T = \int_0^x \bar{F}(u)du + \int_x^\infty \bar{F}(u)du, \ x \geq 0,$$

show that $\mathsf{P}(T^0 > x) \geq \bar{F}(x)$, *that is,* $T^0 \geq_{st} T$, *and in particular,*

$$\mathsf{E}T \leq \mathsf{E}T^0 = \frac{\mathsf{E}T^2}{\mathsf{E}T}. \tag{2.5.34}$$

Remark 2.8 The statement of Problem 2.14 holds for any renewal process, and in particular for the interarrival time $\tau^0(t) := \tau(t) + \widehat{\tau}(t)$ covering instant t. The distribution (2.5.33) is also called the *spread* (tail) distribution of the original distribution F [15]. Note that (2.5.34), as well as Pollaczek-Khintchine formula (2.5.20) and formula (2.5.22), reflect the so-called *inspection paradox* from renewal theory [2]. This paradox expresses a bias caused by the fact that *large values of time t are covered by larger renewal intervals.*

The following result generalizes (1.2.17) (see Theorem 10.2 in [16] and also [17]).

Problem 2.15 *Prove that, if* $\mathsf{E}T^r < \infty$ *for some* $r > 0$, *then w.p.1,*

$$\lim_{t\to\infty} \frac{T(t)}{t^{1/r}} = 0. \tag{2.5.35}$$

Hint: rewrite (2.5.35) as $\lim_{t\to\infty} T^r(t)/t = 0$, *denote by* $T_n^r = \zeta_n$, *and, noting that*

$$T^r(t) \leq \max_{1\leq n\leq N(t)} \zeta_n,$$

where $N(t)$ is the number of regenerations in $[0, t]$ (see (1.2.8)), apply the proof following formulas (2.3.11)–(2.3.14).

Remark 2.9 It is insightful to note that condition $\mathsf{E}T < \infty$ implying $T_1 < \infty$ w.p.1, is not confined to the zero-delayed case where $T_1 =_{st} T$. For instance, consider a single-server $GI/G/1$ system with initial state $\nu_1 = N \geq 1$ and $S(0^-) = 0$, and with a *reverse* Last-In-First-Out (LIFO) service discipline, in which the server selects the latest customer arrival to be served next. This service discipline does not alter the distribution of a busy period or the remaining amount of work observed by an arrival, compared to FIFO. Then it is easy to deduce that the length of the first regeneration period is distributed as a w.p.1 finite random sum

$$T_1 =_{st} B_1 + \cdots + B_{N+1} + I_0 \,, \tag{2.5.36}$$

where $\{B_i\}$ are iid copies of the 'standard' busy period B, and I_0 is a 'standard' idle period, in the zero initial state FIFO system. Indeed, under LIFO, each of $N + 1$ customers generates a busy period B, and the initial queue is then 'exhausted' after $N + 1$ standard busy periods, after which the first regeneration point appears when the first idle period ends. If $S(0^-) = x > 0$, then the (oldest) customer being served at instant $t = 0$ can continue service only if there are no other customers in the system. In this case, the regeneration point occurs when the sum of the standard idle periods exceeds the level x. This approach can also be extended to some multiserver systems, for instance, using monotonicity properties (see Sect. 3.3, Chap. 3) and a *random assignment* system, in which each arrival independently selects server i with probability $1/m$, leading to a set of m single-server FIFO queueing systems, see [18]. A modification of the random assignment system in case of non-identical servers is considered in Sect. 4.4.

Problem 2.16 *Use the LIFO scheduling rule to obtain the following expression for the distribution function G of the busy period B in a positive recurrent $M/G/1$ queuieng system with input rate λ and service time S with distribution function F:*

$$G(x) = \mathsf{P}(B \leq x) = \int_0^x \sum_{n=0}^{\infty} e^{-\lambda u} \frac{\lambda u}{n!} G^{(n)}(x-u) dF(u) \,, \quad x \geq 0 \,, \tag{2.5.37}$$

where $G^{(n)}$ represents the n-fold convolution of the function G with itself ($G^{(0)} := 1$). In addition, apply the Laplace-Stieltjes transform (LST) to both sides of (2.5.37) to obtain the functional equation

$$g(z) = f(z + \lambda - \lambda g(z)) \,, \quad z \geq 0 \,,$$

relating the LST of the busy period $g(z) = \mathsf{E}e^{-zB}$ and the LST of the service time $f(z) = \mathsf{E}e^{-zS}$. (See Chap. 4 in [19].)

2.6 Notes

The stability conditions derived in this chapter are well-known, and were included in order to highlight various aspects related to the regenerative method based on classical regenerations that we present in this book. The regeneration condition $\mathsf{P}(\tau > S) > 0$ is not an important limitation from a practical point-of-view, however it is still possible to develop the stability analysis of the $GI/G/m$ system under condition $\rho < m$ only, which in general only implies the (weaker 'regeneration') condition $\mathsf{P}(\tau > S/m) > 0$. Under this condition, the basic Kiefer-Wolfowitz Markov chain (2.5.27) turns out to be Harris positive recurrent, see the analysis in [2] (Chapter XII) which also covers a continuous-time setting. In this case the 'unloading' procedure leading to *one-dependent* regeneration becomes highly complicated (see also Sect. 4.6, Chap. 10 in [20]). We would also like to point out that the stability analysis of this process goes back to the famous paper [14]. The regeneration condition $\mathsf{P}(\tau > S) > 0$ is by no means new, it suffices to mention paper [21] (Theorem 2.2). The regeneration condition ensures the return of the stochastic process to a regeneration state. In our analysis, as a rule, the regeneration condition turns out to be rather intuitive, but in general it may take a variety of forms, and a considerable effort could be required to check that, under this condition, a regeneration occurs in a finite time period. An important particular case is that of an unbounded interarrival time, in which case the regeneration condition is satisfied automatically. As to instability of the multiserver system, we also mention the paper [22], and the related work [23]. We note that the analysis in (2.3.2)–(2.3.6) is in part adopted from [24]. Also it is useful to note that the popular fluid stability analysis method does not work in the boundary case $\rho = m$. (For the reader interested in fluid stability analysis, we recommend the paper [25] and monograph [26].) An approach based on a decomposition of the regenerative process on the hierarchically embedded regeneration cycles, in order to simplify the calculation of the performance measures, is summarized in the survey paper [27].

It follows from (2.3.21) that, for a regenerative process $\{X_n\}$ with regeneration instants $\{\theta_k\}$ satisfying the corresponding regeneration condition,

$$\phi_x(\mathbb{B}) := \mathsf{E}_x\Big(\sum_{k=1}^{\theta_1} \mathsf{1}(X_k \in \mathbb{B})\Big) \le C\,, \tag{2.6.1}$$

where the constant C depends on the (bounded) set $\mathbb{B}$ but not on the initial state x. Denote by φ the distribution of X_{θ_k}. It then follows from (2.6.1) that the mean number of visits to the set $\mathbb{B}$ during a (generic) regeneration period,

$$\phi_\varphi(\mathbb{B}) = \int \phi_x(\mathbb{B})\varphi(dx)\,, \tag{2.6.2}$$

is finite *irrespective* of whether or not the process is positive recurrent. This is evident when the process is positive recurrent because $\phi_\varphi(\mathbb{B}) \le \mathsf{E}_\varphi\,\theta < \infty$, while if

$\mathsf{E}_\varphi\theta = \infty$, then using a proof by contradiction we can show that $\lim_{n\to\infty} \mathsf{P}_x(X_n \in \mathbb{B}) = 0$ for any x. A similar property holds as well (if $\mathsf{P}_\varphi(\mathbb{B}) < \infty$) for a *null-recurrent* Harris Markov chain having, by definition, an infinite mean distance between (non-classical) regenerations (see Corollary 3.9 in Chapter VII, [2]). Therefore, the method that we present in this book can be generally applied for the stability analysis of systems with non-classical regenerations, as long as the regeneration instants constitute a renewal process. In this regard, also see [28, 29].

The inequalities (2.5.30) are based on the results from [30]. The theory of cumulative processes originally developed in [17] contains the requirement of so-called 'bounded variation' of a *variation process* $\int_0^t |df(X(u))|$ associated with the original cumulative process (1.2.12). Using (2.5.30), then it is easy to establish that all variation processes associated with the regenerative processes that we consider have w.p.1 bounded variation.

References

1. Billingsley, P.: Probability and Measure, 3rd edn. Wiley, New York (1995)
2. Asmussen, S.: Applied Probability and Queues, 2nd edn. Springer-Verlag, New York (2003)
3. Morozov, E.: The stability of non-homogeneous queueing system with regenerative input. J. Math. Sci. **89**, 407–421 (1997)
4. Shiryaev, A.: Probability. Springer-Verlag, New York (1996)
5. Sigman, K., Thorisson, H., Wolff, R.W.: A note on the existence of regeneration times. J. Appl. Prob. **31**(4), 1116–1122 (1994)
6. Feller, W.: An Introduction to Probability Theory and its Applications, II, 2nd edn. Wiley, New York (1971)
7. Murphree, M., Smith, W.: On transient regenerative processes. J. Appl. Prob. **23**, 52–70 (1986)
8. Wolff, R.W.: Poisson arrivals see time averages. Oper. Res. **30**(2), 223–231 (1982)
9. Kaj, I.: Stochastic modeling in broadband communications systems. SIAM monographs on mathematical modeling and computation (2002)
10. Cooper, R.B., Niu, S.-C., Srinivasan, M.M.: Some reflections on the renewal-theory paradox in queueing theory. Festschrift in honor of Professor Ryszard Syski. Special Issue J. Appl. Math. Stoch. Anal. 11(3), 355–368 (1998)
11. Müller, A., Stoyan, D.: Comparisons Methods for Stochastic Models and Risks. Wiley and Sons, Hoboken (N-J) (2002)
12. König, D., Schmidt, V.: Stochastic inequalities between customer-stationary and time-stationary characteristics of queueing systems with point processes. J. Appl. Prob. **17**(3), 768–777 (1980)
13. Morozov, E., Rumyantsev, A., Kalinina, K.: Inequalities for workload process in queues with NBU/NWU input. Adv. Intel. Syst. Comp. **659**, 535–544 (2017)
14. Kiefer, J., Wolfowitz, J.: On the theory of queues with many servers. Trans. Amer. Math. Soc. **78**, 1–18 (1955)
15. Sigman, K.: Stationary Marked Point Processes. In: Springer Handbook of Engineering Statistics, Hoang Pham (Ed.), Chapter 8, 137-152, Springer-Verlag, London (2006)
16. Gut, A.: Stopped Random Walks Limit Theorems and Applications (2nd Edition). Springer Science+Business Media (2009)
17. Smith, W.L.: Regenerative stochastic processes. Proc. Royal Soc. (Series A) **232**, 6–31 (1955)
18. Sigman, K.: Exact simulation of the stationary distribution of the FIFO $M/G/c$ queue: the general case for $\rho<c$. Queueing Syst. **70**, 37–43 (2012)

19. Gnedenko, B., Kovalenko, I.N.: An Introduction to Queueing Theory, 2nd edn. Birkhäuser, Basel (1989)
20. Thorisson, H.: Coupling, Stationarity, and Regeneration. Springer-Verlag, New York, Probability and its Applications (2000)
21. Whitt, W.: Embedded renewal process in the $GI/G/s$ queues. J. Appl. Prob. **9**, 650–658 (1972)
22. Morozov E., (2002). Instability of nonhomogeneous queueing networks
23. El-Taha, M.: Pathwise rate-stability for input-output processes. Queueing Syst. **22**, 47–63 (1996)
24. Kaspi, H., Mandelbaum, A.: Regenerative closed queueing networks. Stoch. Stoch, Rep. **39**(4), 239–258 (1992)
25. Dai, J.: On positive Harris recurrence of multiclass queueing networks: a unified approach via fluid limit models. Ann. Appl. Prob. **5**, 49–77 (1995)
26. Chen, H., Yao, D.D. (eds.): Fundamentals of Queueing Networks: Performance, Asymptotics, and Optimization. Springer, New York (2001)
27. Rykov, V.V.: Decomposable semi-regenerative processes: review of theory and applications. Reliability Theor. Appl. **16**(2) (2021)
28. Morozov, E.V.: Wide sense regenerative processes with applications to multi-channel queues and networks. Acta. Appl. Math. **34**, 189–212 (1994)
29. Morozov, E.: Weak regeneration in modeling of queueing processes. Queueing Syst. **46**, 295–315 (2004)
30. Takács, L.: The limiting distribution of the virtual waiting time and the queue size for a single-server queue with recurrent input and general service times. Sankhyā: Indian J. Stat. (Series A) **25**(1), 91–100 (1963)

Chapter 3
Tightness and Monotonicity

As became clear throughout the analysis presented in the previous chapter, the tightness property of stochastic processes plays an important role in the regenerative stability analysis of queuieng processes. Another topic that is important for the stability analysis presented in this book, is the monotonicity of the relevant processes in the queueing systems that are considered.

For this reason, in this chapter, we focus on the tightness property, and also present simple proofs of some monotonicity results for multiserver systems, that hopefully help the reader to gain some insight into this issue.

3.1 Tightness of the Queueing Processes

Consider a real-valued d-dimensional *non-negative* stochastic process, denoted by

$$\mathbf{X}(t) = (X_1(t), \ldots, X_d(t)),\ t \geq 0\,,$$

which is sufficient for the purposes in this book. The process $\{\mathbf{X}(t)\}$ is said to be *tight* if, for any $\varepsilon > 0$, there exists a bounded set $\mathbb{B} = [0,\ B] \times \cdots \times [0,\ B]$ (d times) such that

$$\inf_{t \geq 0} \mathsf{P}(\mathbf{X}(t) \in \mathbb{B}) \geq 1 - \varepsilon\,. \tag{3.1.1}$$

(This is a particular case of a more general definition, see for instance Chapter III in [1].) Evidently, if (3.1.1) holds, then each component process $\{X_i(t)\}$ is tight, and vice versa: if the processes $\{X_i(t)\}$, $i = 1, \ldots, d$, are tight then the process $\{\mathbf{X}(t)\}$ is tight as well. Also, if the process $\{\mathbf{X}(t)\}$ is tight, then in particular, the process $\left\{\sum_{i=1}^{d} X_i(t)\right\}$ is tight.

E. Morozov and B. Steyaert, *Stability Analysis of Regenerative Queueing Models*,
https://doi.org/10.1007/978-3-030-82438-9_3

Problem 3.1 Prove the tightness of the processes $\{\mathbf{X}(t) := (X_1(t),\ X_2(t))\}$ and $\{X_1(t) + X_2(t)\}$ if the processes $\{X_1(t)\}$, $\{X_2(t)\}$ are tight.

In our analysis, we often refer to the tightness property of the remaining renewal time process, and in particular, of the remaining interarrival time process $\{\tau(t)\}$ defined in (2.1.14). Indeed, if $\mathsf{E}\tau < \infty$ and the interval τ is non-lattice, then $\tau(t) \Rightarrow \tau_e$. Note that the random variable τ_e is *proper*, i.e., $\tau_e < \infty$ w.p.1, because of $\mathsf{E}\tau < \infty$, see Remark 3.1 below. Otherwise, if τ is *lattice with a span* $d > 0$, then $\tau(nd) \Rightarrow \tau_e,\ n \to \infty$. It is easy to prove in both cases that the process $\{\tau(t)\}$ is tight. For instance, for the non-lattice case, for any $\varepsilon > 0$, select a value x_ε such that $\mathsf{P}(\tau_e \le x_\varepsilon) \ge 1 - \varepsilon/4$ (this is possible since τ_e is proper), and since

$$\lim_{t\to\infty} \mathsf{P}(\tau(t) \le x_\varepsilon) = \mathsf{P}(\tau_e \le x_\varepsilon), \tag{3.1.2}$$

then choose t_ε such that

$$\inf_{t\ge t_\varepsilon} \mathsf{P}(\tau(t) \le x_\varepsilon) \ge \mathsf{P}(\tau_e \le x_\varepsilon) - \varepsilon/4 \ge 1 - \varepsilon/2\,.$$

On the other hand, because $\tau(t) \le t_\varepsilon + \tau(t_\varepsilon)$ for $t \le t_\varepsilon$, then

$$\begin{aligned}\mathsf{P}(\tau(t) \le t_\varepsilon + x_\varepsilon) &\ge \mathsf{P}\Big(\tau(t) \le t_\varepsilon + \tau(t_\varepsilon),\ \tau(t_\varepsilon) \le x_\varepsilon\Big)\\ &\ge \mathsf{P}\big(\tau(t) \le t_\varepsilon + \tau(t_\varepsilon)\big) - \mathsf{P}(\tau(t_\varepsilon) > x_\varepsilon)\\ &\ge 1 - \varepsilon/2, \quad \text{for } t \le t_\varepsilon\,,\end{aligned} \tag{3.1.3}$$

and it then follows that

$$\inf_{t\ge 0} \mathsf{P}(\tau(t) \le x_\varepsilon + t_\varepsilon) \ge 1 - \varepsilon/2\,. \tag{3.1.4}$$

Since ε was arbitrarily chosen, inequality (3.1.4) implies the tightness of the remaining renewal time.

Remark 3.1 It is important to note that the convergence $\tau(t) \Rightarrow \tau_e$ means that (3.1.2) holds for any point x in which the limiting function $\mathsf{P}(\tau_e \le x)$ is *continuous* [2]. But as it follows from (2.3.3), this function can also be written as

$$\mathsf{P}(\tau_e \le x) = \frac{1}{\mathsf{E}\tau}\int_0^x \mathsf{P}(\tau > u)du\,,$$

and therefore is continuous at each point $x \ge 0$, and also $\mathsf{P}(\tau_e < \infty) = 1$.

Problem 3.2 Prove the tightness of $\{\tau(t)\}$ for the lattice case.

As a basic example, consider the $GI/G/m$ queueing system, with FCFS service discipline, renewal input with iid interarrival times $\tau_n = t_{n+1} - t_n$, and input rate

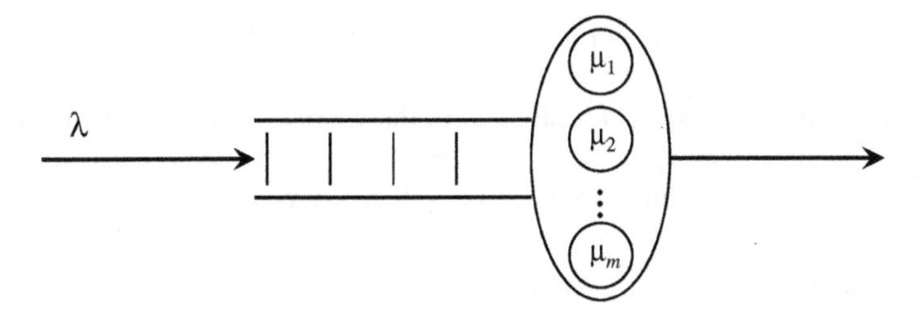

Fig. 3.1 The $GI/G/m$ queueing system with non-equivalent servers

$\lambda = 1/\,\mathsf{E}\tau$ (see Fig. 3.1). Assume that the servers in general are non-identical, and let $\{S_n^{(i)}\}$ represent the iid service times assigned to the consecutive customers served by server i, with rate $\mu_i = 1/\,\mathsf{E}S^{(i)},\ i = 1, \ldots, m$. We assume throughout this section that the following regeneration condition holds:

$$\min_{1 \le i \le m} \mathsf{P}(\tau > S^{(i)}) > 0\,. \tag{3.1.5}$$

If the system is positive recurrent, then the remaining regeneration time process $\{T(t)\}$ is tight, and the total amount of the remaining work (which includes the remaining service times, if any) is a tight process as well, see the proof of Theorem 2.4. Relying on this argument, one can also establish the tightness of other processes describing the behaviour of the positive recurrent queueing system.

Remark 3.2 In the discussion above, the workload process can be replaced by the queue-size process (keeping track of the number of customers in the system) as well, see again the proof of Theorem 2.4.

It follows from analysis developed in Sect. 2.4 that if the system is *not positive recurrent*, i.e., $\mathsf{E}T = \infty$, then the basic processes cannot be tight. However, as we will show, the attained and remaining service time processes remain tight even in the *non-positive recurrent* case.

To prove this, we consider an arbitrary (fixed) server i, and, in order to simplify our notation, *omit the server index* i. To avoid trivial complications, we also assume throughout this subsection a zero initial state.

Now, building on the basic service times sequence $\{S_n\}$, we construct the renewal process

$$\widehat{Z}_n = S_1 + \cdots + S_n\,,\ n \ge 1\,,$$

and denote the associated *remaining* and *attained* renewal times at instant t by $\widehat{S}(t)$ and $a(t)$ respectively, for instance, see (2.5.15), (2.5.31). Then in view of (3.1.4), the process $\{\widehat{S}(t)\}$ is tight. On the other hand, it is easy to check that the following relation holds for each $x \ge 0$:

$$\{\widehat{S}(t) > x\} = \{a(t+x) \geq x\}\ ,\ t \geq 0\ . \tag{3.1.6}$$

Due to the tightness of $\{\widehat{S}(t)\}$, for any $\varepsilon > 0$, there exists a value x_ε such that

$$\inf_{t \geq 0} \mathsf{P}(\widehat{S}(t) > x) \leq \varepsilon\ ,\ \ x \geq x_\varepsilon\ ,$$

and consequently, from (3.1.6), we find

$$\inf_{t \geq 0} \mathsf{P}(a(t+x_\varepsilon) \geq x_\varepsilon) = \inf_{t \geq x_\varepsilon} \mathsf{P}(a(t) \geq x_\varepsilon) \leq \varepsilon\ . \tag{3.1.7}$$

Because $\mathsf{P}(a(t) \geq x_\varepsilon) = 0$ if $t < x_\varepsilon$, then it follows from (3.1.7) that indeed

$$\inf_{t \geq 0} \mathsf{P}(a(t) \geq x) \leq \varepsilon\ ,\ \ \ x \geq x_\varepsilon\ .$$

Since ε is arbitrary, we conclude that the process $\{a(t),\ t \geq 0\}$ is tight as well. For further use, it is convenient to formulate these results (and (3.1.4) as well) as the following statement.

Theorem 3.1 *If the mean renewal time is finite, then the remaining and attained renewal time processes are tight.*

Now we return to the attained and remaining service time in the real service process. The main difficulty in the analysis of the tightness of these processes lies in the presence of the idle periods. These idle periods are random, and a delicate construction (proposed in [3]) is required to establish a correspondence between the renewal process and the actual service time process.

Let $S^*(t)$ be the *attained service time* at instant t, and we set $S^*(t) = 0$ if $\nu(t) = 0$.

Theorem 3.2 *The process $\{S^*(t),\ t \geq 0\}$ is tight.*

Proof We maintain the original successive service times, by copying them from the basic sequence $\{S_n\}$ until the first idle period of the server appears. Then we omit the service times which start within an *idle period of the server* and, when the server becomes busy again at some instant z, select as the next service time the first renewal interval in the renewal process $\{\widehat{Z}_n\}$ which starts *after instant z*. We continue this procedure until the second idle period appears, and keep repeating it for consecutive idle periods. From the iid *omitted* intervals, denoted by $\{\widetilde{S}_n\}$, we then construct the renewal process,

$$\widetilde{Z}_n = \widetilde{S}_1 + \cdots + \widetilde{S}_n,\ \ n \geq 1\ ,$$

with the remaining renewal time $\widetilde{S}(t)$ at instant t. Note that the processes $\{\widehat{Z}_n\}$ and $\{\widetilde{Z}_n\}$ are equivalent and independent, and that the process $\{\widetilde{S}(t)\}$ is tight. Due to this construction, the corresponding interval in the renewal process $\{\widehat{Z}_n\}$ *outruns* the same (actual) service time, and we denote by $h(t) \geq 0$ the *distance* between these *identical intervals* at time instant t. We set $h(t) = 0$ if the server is idle at instant t, i.e., if $S(t) = 0$. Also, let

$$I(t) = \int_0^t \mathsf{1}(\nu(u) = 0)du \ ,$$

be the idle time of the server in the interval $[0,\ t]$. It was proved in [4] that if $h(t) > 0$, then the following equality holds:

$$h(t) = \widetilde{S}(I(t)) = \min_{n\geq 1}(\widetilde{Z}_n - I(t) : \widetilde{Z}_n > I(t)) \ , \tag{3.1.8}$$

expressing that the distance $h(t)$ equals the remaining renewal time at instant $I(t)$ in the process composed by the omitted service times. Therefore, in general,

$$h(t) \leq \widetilde{S}(I(t)) \ , \quad t \geq 0 \ . \tag{3.1.9}$$

Figure 3.2 provides an intuitive explanation of equality (3.1.8): at any instant $t \in [\widehat{Z}_4,\ \widehat{Z}_6]$ a renewal interval in the process $\{\widehat{Z}_n\}$ outruns the *same real service time* by a value $h(t) = \widetilde{S}(I(t))$. By construction, conditioned on the event $\{I(t) = x\}$, the random variable $\widetilde{S}(I(t))$ is distributed as the remaining renewal time $\widetilde{S}(x)$. Therefore, it follows from (3.1.9) that, for any $t, x \geq 0$,

$$\begin{aligned} \mathsf{P}(h(t) \leq x) &\geq \int_0^t \mathsf{P}(\widetilde{S}(u) \leq x)\ \mathsf{P}(I(t) \in du) \\ &\geq \inf_{u\geq 0} \mathsf{P}(\widetilde{S}(u) \leq x) \ . \end{aligned} \tag{3.1.10}$$

By Theorem 3.1 the remaining renewal time process $\{\widetilde{S}(t)\}$ is tight, and then the process $\{h(t)\}$ is tight as well. It is easily seen that if a time instant t is covered by the *same interval* in the renewal sequence $\{\widehat{Z}_n\}$ and in the actual service process, then $S^*(t) = h(t) + a(t)$, otherwise $S^*(t) \leq h(t)$, implying

$$S^*(t) \leq h(t) + a(t) \ , \quad t \geq 0 \ , \tag{3.1.11}$$

and in view of the tightness of $\{a(t)\}$ and $\{h(t)\}$, the tightness of the process $\{S^*(t)\}$ follows. □

Problem 3.3 Using (3.1.11), prove the tightness of $\{S^*(t)\}$ in detail.

Theorem 3.3 *The remaining service time process* $\{S(t),\ t \geq 0\}$ *is tight.*

Proof By analogy with (3.1.6),

$$\{S(t) > x\} = \{S^*(t + x) \geq x\} \ , \quad x \geq 0 \ , \tag{3.1.12}$$

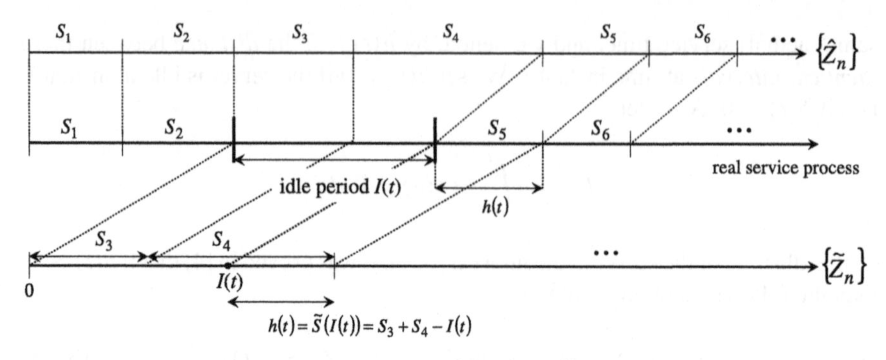

Fig. 3.2 The relation between the renewal processes $\{\widehat{Z}_n\}$, $\{\widetilde{Z}_n\}$ and the real service process. In this example: the 'distance' at instant $t \in [\widehat{Z}_4, \widehat{Z}_6]$ between the real service process and $\{\widehat{Z}_n\}$ equals $h(t) = \widetilde{S}(I(t)) = S_3 + S_4 - I(t) > 0$

and, by the tightness of $\{S^*(t)\}$, we find that,

$$\sup_{t \geq 0} \mathsf{P}(S^*(t+x) \geq x) = \sup_{t \geq 0} \mathsf{P}(S^*(t) \geq x) \to 0\,, \quad x \to \infty\,,$$

and the tightness of $\{S(t)\}$ immediately follows from (3.1.12). □

Remark 3.3 It is worth mentioning that the above proofs indeed do not depend on whether or not the underlying queueing system is positive recurrent. Moreover, it is evident that the continuous-time parameter t can be replaced by a series of deterministic time instants, or an embedded random time sequence, provided these are independent of the service process. In many important cases however, for instance when the embedded random times represent arrival instants, there may exist a dependence between these instants and the service process.

Remark 3.4 Suppose for a moment that we form the real service process by directly using the basic sequence $\{S_n\}$. Then $S(t) = \widehat{S}(t - I(t)$ if $\nu(t) > 0$, implying

$$S(t) \leq \widehat{S}(t - I(t))\,, \quad t \geq 0\,. \tag{3.1.13}$$

Since the process $\{\widehat{S}(t)\}$ is tight, then it might be tempting to deduce the tightness of $\{S(t)\}$ directly from inequality (3.1.13). However, this is not a valid argument, since the time instant $t - I(t)$ is random, and $\widehat{S}(t - I(t))$ depends, via the attained service time $a(t - I(t))$, on the service times starting *before* the instant $t - I(t)$. (An example illustrating this effect can be found in [5].)

3.2 Tightness of the Range of the Workload Components

It has been proved in [6] that, in the $GI/G/m$ system with *identical servers*, renewal input, a FCFS service discipline and a *zero initial state*, the difference between the largest and the smallest remaining work components (called the *range*) is a tight process. Our purpose below is to extend this result to a more general setting.

Consider a queueing system with $m \geq 2$ servers, arbitrary arrival instants $0 = t_1 < t_2 < \ldots$, and independent iid sequences of service times $\{S_n^{(i)}\}$ intended for each server $i = 1, \ldots, m$. Assume that $\mathsf{1}_n^{(i)} = 1$ if the n-th customer is served by server i (and $\mathsf{1}_n^{(i)} = 0$ otherwise), then his service time is given by

$$S_n = \sum_{i=1}^{m} \mathsf{1}_n^{(i)} S_n^{(i)}, \quad n \geq 1 .$$

Let

$$\mathbf{W}_n = (W_n^{(1)}, \ldots, W_n^{(m)}) ,$$

be the *Kiefer-Wolfowitz workload vector* satisfying recursion (2.5.27). Since $W_n^{(m)}$ and $W_n^{(1)}$ respectively are the largest and smallest workload components, we can introduce the non-negative workload *range* process as

$$\Delta_n = W_n^{(m)} - W_n^{(1)} , \quad n \geq 1 .$$

The following result is an extension of the 'essential' lemma in [6] to an m-server system with *non-identical servers, arbitrary input and any initial state.*

Theorem 3.4 *The range process $\{\Delta_n\}$ is tight for any initial state $\mathbf{W}_1 = \mathbf{x}_1 = (x_1, \ldots, x_m)$.*

Proof Following [6], denote by

$$\mathsf{B}_n = W_n^{(m)}(m-1) - \sum_{i=1}^{m-1} W_n^{(i)}, \quad n \geq 1 .$$

Obviously,

$$\Delta_1 \leq \mathsf{B}_1 = x_m(m-1) - \sum_{i=1}^{m-1} x_i ,$$

and it is easy to check the following inequalities

$$\mathsf{B}_{n+1} \leq \mathsf{B}_n - S_n , \quad \text{if } S_n \leq \Delta_n ,$$
$$\mathsf{B}_{n+1} \leq (m-1)S_n , \quad \text{if } S_n \geq \Delta_n, \ n \geq 1 ,$$

implying

$$\mathsf{B}_{n+1} \le \max\Big((m-1)S_n,\ \mathsf{B}_n - S_n\Big),\quad n \ge 1\,.$$

Then we recursively obtain the following inequality:

$$\begin{aligned}\mathsf{B}_{n+1} \le \max\Big(&(m-1)S_n,\ (m-1)S_{n-1} - S_n, \ldots, (m-1)S_1 \\ &- S_2 - \cdots - S_n,\ \mathsf{B}_1 - S_1 - \cdots - S_n\Big) =: Y_n,\quad n \ge 1\,,\end{aligned} \tag{3.2.1}$$

where $S_0 := 0$. Since

$$0 \le \Delta_n \le \mathsf{B}_n \le Y_{n-1}\,,$$

it is sufficient to show that the sequence $\{Y_n\}$ is tight. Denote by

$$U_n = \max_{1\le i\le m} S_n^{(i)}\,,\quad V_n = \min_{1\le i\le m} S_n^{(i)}\,,\quad n \ge 1\,,$$

and note that $\{U_n\}$ and $\{V_n\}$ are independent iid sequences. Since $V_n \le S_n \le U_n$, it then follows from (3.2.1) that, for all $n \ge 1$,

$$\begin{aligned}Y_n &= \max\Big(\mathsf{B}_1 - \sum_{k=1}^{n} S_k,\ (m-1)S_k - \sum_{i=k+1}^{n} S_i,\ k=1,\ldots,n\Big) \\ &\le \max\Big(\mathsf{B}_1 - \sum_{k=1}^{n} V_k,\ (m-1)U_k - \sum_{i=k+1}^{n} V_i,\ k=1,\ldots,n\Big) \\ &=_{st} \max\Big(\mathsf{B}_1 - \sum_{k=1}^{n} V_k,\ (m-1)U_k - \sum_{i=1}^{k-1} V_i,\ k=1,\ldots,n\Big) \qquad (3.2.2) \\ &\le \mathsf{B}_1 + \max_{1\le k<\infty}\Big((m-1)U_k - \sum_{i=1}^{k-1} V_i\Big)\,. \qquad (3.2.3)\end{aligned}$$

In the stochastic equality in (3.2.2), we replace the sums $\sum_{i=k+1}^{n} V_i$ by the sums $\sum_{i=1}^{k-1} V_i$, relying on the independence, for each k, between U_k and V_i, $i \ne k$. It is easy to verify that this replacement leaves the distribution of $\max\{\cdot\}$ unchanged. To illustrate this, let $n = 4$, then

$$\Big\{\sum_{i=k+1}^{n} V_i,\ k=1,\ldots,n\Big\} = \Big\{V_2 + V_3 + V_4,\ V_3 + V_4,\ V_4\Big\}\,,$$

while

$$\Big\{\sum_{i=1}^{k-1} V_i,\ k=1,\ldots,n\Big\} = \Big\{V_1,\ V_1 + V_2,\ V_1 + V_2 + V_3\Big\}\,,$$

and, for instance, $(m-1)U_1 - (V_2 + V_3 + V_4)$ and $(m-1)U_4 - (V_1 + V_2 + V_3)$ have the same distribution. Therefore, invoking the SLLN, we find that

$$\lim_{k\to\infty} \frac{1}{k}\sum_{i=1}^{k-1} V_i = \mathsf{E}V > 0\,, \tag{3.2.4}$$

implying, w.p.1,

$$\sum_{i=1}^{k-1} V_i \to \infty,\ k \to \infty\,,$$

and for this reason, also

$$(m-1)U_k - \sum_{i=1}^{k-1} V_i \to -\infty,\ k \to \infty\,.$$

Evidently, this implies

$$\max_{1\le k<\infty}\left((m-1)U_k - \sum_{i=1}^{k-1} V_i\right) < \infty \text{ w.p.1}\,.$$

Since the upper bound (3.2.3) is independent of n, the sequence $\{Y_n\}$ is indeed tight. □

Remark 3.5 We would like to point out that the structure of the input process does not play a role in the proof of Theorem 3.4 and in particular, its statement holds for the system with regenerative input process, see Sect. 4.6. The natural assumption that we implicitly use in this proof (and in the proofs of the following Theorems 3.5 and 3.6) is that $\min_i \mathsf{E}S^{(i)} > 0$, see (3.2.4). (Also see a comment in Sect. 3.4, where we further elaborate on this independence property.)

The statement of Theorem 3.4 can be extended to the continuous-time workload process

$$\mathbf{W}(t) = \Big\{W_1(t), \ldots, W_m(t)\Big\}\,,\ t \ge 0\,,$$

where $W_i(t)$ is the remaining work in *server i* at time t. We denote $n(t) = \min\{n : t_n \ge t\}$, implying $n(t_n) = n$, and define the remaining work vector

$$\mathbf{W}_{n(t)} = \Big\{W^{(1)}_{n(t)}, \ldots, W^{(m)}_{n(t)}\Big\}\,,$$

which is observed by the $1st$ arrival in $[t,\ \infty)$. Define also the range process

$$\Delta_{n(t)} = \max_{i,j} \mid W^{(i)}_{n(t)} - W^{(j)}_{n(t)} \mid \ , \ t \geq 0 \, .$$

A detailed proof of the following statement can be found in [5].

Theorem 3.5 *The range process* $\{\Delta_{n(t)},\ t \geq 0\}$ *is tight.*

Proof The proof is similar to the proof of Theorem 3.4, by replacing index n by $n(t)$. Indeed, in this case

$$U_{n(t)} =_{st} U, \ \ V_{n(t)} =_{st} V \, ,$$

and, for each t and i, the random variable $U_{n(t)-i}$ is independent of V_k for all $k \neq n(t) - i$. Consequently, the remainder of the proof repeats the previous one. □

The next result is a direct corollary of Theorem 3.5.

Theorem 3.6 *Assume that the remaining interarrival time process* $\{\tau(t)\}$ *is tight. Then the range process*

$$\Delta(t) := \max_{i,j} |W_i(t) - W_j(t)|, \ t \geq 0 \, ,$$

is tight.

Proof Since $\Delta(t_n) = \Delta_n$, then we assume that $t \neq t_n$.
(i) Assume that $\tau(t) \leq \min_{1 \leq i \leq m} W_i(t)$, then

$$W_i(t) = W^{(i)}_{n(t)} + \tau(t) \, ,$$

and it follows from the equalities

$$W_i(t) - W_j(t) = W^{(i)}_{n(t)} - W^{(j)}_{n(t)} \, , \ i, j = 1, \ldots, m \, ,$$

that $\Delta(t) = \Delta_{n(t)},\ t \geq 0$.
(ii) Assume that

$$\min_{1 \leq i \leq m} W_i(t) \leq \tau(t) \leq \max_{1 \leq i \leq m} W_i(t) \, ,$$

then, it is easy to verify that $\Delta(t) \leq \Delta_{n(t)} + \tau(t)$. Now we can refer to the tightness of the processes $\{\Delta_{n(t)}\}$, $\{\tau(t)\}$ and to (3.1.11) to complete this part.
(iii) Finally, let

$$\tau(t) \geq \max_{1 \leq i \leq m} W_i(t),$$

then $\Delta(t) \leq \tau(t)$, and the proof is completed. □

In particular, it follows from Theorem 3.6 that the components of workload process vector satisfy the following *solidarity property*:

$$W_1(t) \not\Rightarrow \infty \text{ implies } W_m(t) \not\Rightarrow \infty . \tag{3.2.5}$$

This observation is essential in the stability analysis of multiserver systems developed in the following chapters.

3.3 Monotonicity of the Multiserver System

We now present simple coupling-based proofs of a few monotonicity properties of the multiserver system, which are widely used for the stability analysis presented throughout this book as well.

Consider the multiserver $GI/G/m$ system as above, denoted by Σ, and let $\widetilde{\Sigma}$ be another (similar) system in which the corresponding variables will be marked by the superscript 'tilde'. Maintaining the previous notations, also define, in system Σ, d_n as the departure instant of the n-th customer, and B_n as the n-th service initiation instant. The relations given below can be treated as *stochastic* for the original variables, or as relations w.p.1 for the *coupled variables* defined on the same probability space. In what follows, we will make use of the next results.

Problem 3.4 Show that the Kiefer-Wolfowitz operator R in (2.5.27) preserves the ordering, i.e., $\mathbf{X} \le \mathbf{Y}$ implies $R\mathbf{X} \le R\mathbf{Y}$.

Problem 3.5 Using induction, prove the following *monotonicity* property of the Kiefer-Wolfowitz vector (2.5.27): assumptions

$$\widetilde{\mathbf{W}}_1 \le \mathbf{W}_1 , \ \widetilde{\tau} \ge \tau , \ \widetilde{S} \le S ,$$

imply

$$\widetilde{\mathbf{W}}_n \le \mathbf{W}_n , \ n \ge 1 . \tag{3.3.1}$$

In particular, the *waiting times* are ordered as $\widetilde{W}_n^{(1)} \le W_n^{(1)}, \ n \ge 1$.

Theorem 3.7 *Assume that*

$$\nu_1 = \widetilde{\nu}_1 = 0 , \ \tau = \widetilde{\tau} , \ S \ge \widetilde{S} , \tag{3.3.2}$$

then

$$\widetilde{\nu}_n \le \nu_n , \ n \ge 1 ; \ \widetilde{\nu}(t) \le \nu(t) , \ t \ge 0 . \tag{3.3.3}$$

Proof Due to the coupling, $t_n = \widetilde{t}_n$, and the number of arrivals in both systems during $[0, t]$ equals $A(t)$. Then it follows from (3.3.1) that $\widetilde{W}_n^{(1)} \leq W_n^{(1)}$, and, in view of assumption (3.3.2), the departure instants of customer n in both systems are ordered as

$$d_n = t_n + W_n^{(1)} + S_n \geq t_n + \widetilde{W}_n^{(1)} + \widetilde{S}_n = \widetilde{d}_n \ , \ n \geq 1 \ . \tag{3.3.4}$$

Hence, the number of departures in the time interval $[0, t]$ in both systems are ordered as

$$\widetilde{D}(t) \geq D(t) \ , \ t \geq 0 \ , \tag{3.3.5}$$

and, in particular, denoting by $D_n = D(t_n^-)$, $\widetilde{D}_n = \widetilde{D}(t_n^-)$ the number of departures in the interval $[0, t_n)$, we may write

$$\widetilde{D}_n \geq D_n \ , \ n \geq 1 \ .$$

It then becomes clear that

$$\begin{aligned} \nu_n := \nu(t_n^-) = A(t_n^-) - D_n &= n - 1 - D_n \geq n - 1 - \widetilde{D}_n \\ &= \widetilde{\nu}_n \ , \ n \geq 1 \ , \end{aligned} \tag{3.3.6}$$

and the first inequality in (3.3.3) indeed holds. Now, using (3.3.5), the second inequality in (3.3.3) follows from

$$\nu(t) = A(t) - D(t) \geq A(t) - \widetilde{D}(t) = \widetilde{\nu}(t) \ , \ t \geq 0 \ .$$

□

Because the attained interarrival time process $\{\widehat{\tau}(t)\}$ defined in (2.5.15) is the same in both systems, then it follows from the monotonicity property of the continuous-time analogue of the Kiefer-Wolfowitz vector (2.5.28) that the following statement holds as well.

Theorem 3.8 *Assume that* $\nu_1 = \widetilde{\nu}_1 = 0$, $\tau = \widetilde{\tau}$ *and* $S \geq \widetilde{S}$, *then*

$$\mathbf{W}(t) \geq \widetilde{\mathbf{W}}(t) \ , \ t \geq 0 \ .$$

In particular, the virtual waiting times are ordered as

$$W_1(t) \geq \widetilde{W}_1(t) \ , \ t \geq 0 \ .$$

Theorem 3.9 *Assume that*

$$\widetilde{\mathbf{W}}_1 \leq \mathbf{W}_1 \ , \ \tau \leq \widetilde{\tau} \ , \ S = \widetilde{S} \ . \tag{3.3.7}$$

Then

$$\widetilde{\nu}_n \le \nu_n \, , \; n \ge 1 \, . \tag{3.3.8}$$

Proof The assumption $\tau \le \widetilde{\tau}$ implies $t_n \le \widetilde{t}_n$, and the difference $\Delta_n := \widetilde{t}_n - t_n \ge 0$ is non-decreasing in n. Relying on (3.3.7) and $\widetilde{W}_n^{(1)} \le W_n^{(1)}$ (see Problem 3.5), the departure instants of customer n in both systems are ordered as (using (3.3.4))

$$\begin{aligned}\widetilde{d}_n &= \widetilde{t}_n + \widetilde{W}_n^{(1)} + S_n = \Delta_n + t_n + \widetilde{W}_n^{(1)} + S_n \\ &\le \Delta_n + t_n + W_n^{(1)} + S_n = \Delta_n + d_n \, , \; n \ge 1 \, ,\end{aligned}$$

and hence, for $n \ge k \ge 1$, the inequalities

$$\widetilde{d}_k \le d_k + \Delta_k \le d_k + \Delta_n \, ,$$

hold. Then

$$\begin{aligned}D_n &= \max(k : d_k < t_n) = \max(k : d_k + \Delta_n < t_n + \Delta_n) \\ &\le \max(k : d_k + \Delta_k < \widetilde{t}_n) \le \max(k : \widetilde{d}_k < \widetilde{t}_n) = \widetilde{D}_n \, .\end{aligned}$$

Consequently, the number of departures in the system $\widetilde{\Sigma}$ during the time interval $[0, \widetilde{t}_n)$ dominates the number of departures in the system Σ during the time interval $[0, t_n)$. Since $\nu_n = \nu(t_n^-)$ and $\widetilde{\nu}_n = \widetilde{\nu}(\widetilde{t}_n^-)$, then (3.3.8) follows as in (3.3.6). □

3.4 Notes

The proof of Theorem 3.2, borrowed from [5], is based on the profound and elegant method developed in [3].

Because the proof of Theorem 3.4 is independent of the structure of the interarrival times, it allows to apply this result in a network context, where the input to a network station typically has a very complicated structure. It may be tempting to simplify the proof of Theorem 3.4, for instance, using the following argument. Consider an arrival time instant t_k when the minimal component of the workload vector becomes the maximal one, i.e.,

$$W_k^{(1)} + S_k \ge W_k^{(m)} \, ,$$

or, equivalently, $\Delta_k \le S_k$. Note however that S_k does not yield a useful upper bound for Δ_k since the service time S_k has an *unknown conditional distribution* on the event $\{S_k \ge \Delta_k\}$.

The monotonicity results given above can also be found in [7–9]. (Paper [9] also contains the revised proofs of some results found in [7].) More monotonicity results for queueing systems are in [10]. A coupling approach has been applied in [11] to establish the monotonicity of a class of queueing systems with *dependent governing variables* (such as interarrival and service times).

References

1. Shiryaev, A.: Probability. Springer-Verlag, New York (1996)
2. Billingsley, P.: Probability and Measure, 3rd edn. Wiley, New York (1995)
3. Borovkov, A.A.: Some limit theorems in the queueing theory II. Theor. Prob. Appl. **10**, 375–400 (1965)
4. Iglehart, D.L., Whitt, W.: Multiple channel queues in heavy traffic I. Adv. Appl. Prob. **2**(1), 150–177 (1970)
5. Morozov, E.: The tightness in the ergodic analysis of regenerative queueing processes. Queueing Syst. **27**, 179–203 (1997)
6. Kiefer, J., Wolfowitz, J.: On the theory of queues with many servers. Trans. Amer. Math. Soc. **78**, 1–18 (1955)
7. Jacobs, D.R., Schach, S.: Stochastic order relationships between $GI/G/k$ queues. Ann. Math. Stat. **43**(5), 1623–1633 (1972)
8. Müller, A., Stoyan, D.: Comparisons Methods for Stochastic Models and Risks. Wiley and Sons, Hoboken (N-J) (2002)
9. Whitt, W.: Comparing counting processes and queues. Adv. Appl. Prob. **13**, 207–220 (1981)
10. Asmussen, S.: Applied Probability and Queues, 2nd edn. Springer-Verlag, New York (2003)
11. O'Brien, G.L.: Inequalities for queues with dependent interarrival and service times. J. Appl. Prob. **12**, 653–656 (1975)

Chapter 4
Generalizations of Multiserver Systems

In this chapter, we study the stability of some important extensions of the classical $GI/G/m$ system: a multiserver system with non-identical servers (that was already in part discussed in Chap. 3), a finite-buffer system, a system with an infinite number of servers, and a system with a regenerative input flow. The multiserver system with non-identical servers is difficult to analyze because it does not possess the monotonicity property and belongs to a special class of state-dependent systems where the service time of a customer depends on the assigned server. The system with regenerative input can be considered as a link between an isolated queue and a regenerative network in which internal streams preserve the regeneration property. (A more detailed analysis of this problem can be found in [1, 2]. We also touch upon this topic in Sect. 10.3.)

4.1 The Multiserver Multiclass System

In this section, we analyze a *multiserver multiclass* system with *class-dependent and server-dependent* service times, shown in Fig. 4.1. This analysis particularly covers a multiserver system with non-identical servers, meaning that a customer's service time depends on the server it is assigned to. A key element of the analysis below is the consideration of the system's operation in a *saturated regime*.

Consider an m-server multiclass queueing system with K classes of customers, generated by a common renewal input process with iid interarrival times $\tau_n = t_{n+1} - t_n$, $n \geq 1$, and arrival rate $\lambda = 1/\mathsf{E}\tau$. It is assumed that each new customer belongs to class k with probability p_k, implying that the input rate of class-k customers is given by $\lambda_k = \lambda p_k$, $k = 1, \ldots, K$. We consider a FCFS work-conserving service discipline; therefore, p_k is also the probability that an arbitrary customer entering any server is a class-k customer. We also consider independent sequences of iid service times $\{S_{ik}^{(n)}, n \geq 1\}$ of class-k customers at server i, with rates

E. Morozov and B. Steyaert, *Stability Analysis of Regenerative Queueing Models*,
https://doi.org/10.1007/978-3-030-82438-9_4

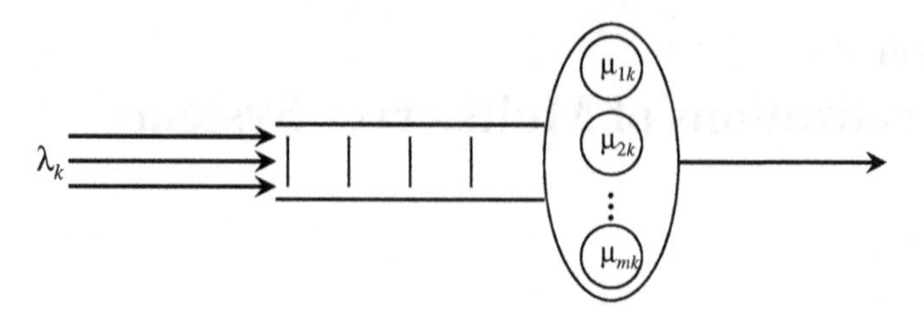

Fig. 4.1 The multiclass $GI/G/m$ system with class- and server-dependent service times

$$\mu_{ik} = 1/\mathsf{E}S_{ik},\ i = 1, \ldots, m;\ \ k = 1, \ldots, K\ .$$

(In the notations adopted throughout this section, sub/superscript i refers to server i, and sub/superscript k to the customer class.) We would like to emphasize that the service times S_{ik} are in general not equivalent for different i and k.

Assume that the binary random variable $\mathsf{1}_{ik}^{(n)} = 1$ if the nth customer entering server i belongs to class k, and $\mathsf{1}_{ik}^{(n)} = 0$ otherwise, $n \geq 1$. (In this notation, index n counts the customers that are intended for server i only.) It then follows that the service time $S_n^{(i)}$ of the nth customer at server i is given by

$$S_n^{(i)} = \sum_{k=1}^{K} \mathsf{1}_{ik}^{(n)} S_{ik}^{(n)}\ ,\ n \geq 1\ . \tag{4.1.1}$$

We emphasize that, although the servers may be different, each customer (entering any server) is of type k with probability p_k, and then for each k, the iid sequences $\{1_{ik}^{(n)},\ n \geq 1\}$ (with generic element $\mathsf{1}_{ik}$) are stochastically equivalent for all i. In particular, for any k we may write

$$\mathsf{E}\mathsf{1}_{ik} = \mathsf{E}\mathsf{1}_{jk} = p_k\ ,\ \ \text{for any } i, j\ .$$

For different i, the service times S_{ik} are in general differently distributed, however, for each i, the random variables $\{S_n^{(i)}\}$ in (4.1.1) are iid with generic element $S^{(i)}$ and expectation

$$\mathsf{E}S^{(i)} = \sum_{k=1}^{K} \frac{p_k}{\mu_{ik}},\ \ i = 1, \ldots, m\ .$$

Let $\nu_i(t)$ be the number of the customers in the system at time instant t which are intended for server i, and let

$$\nu(t) = \sum_{i=1}^{m} \nu_i(t),\ t \geq 0\ ,\ \nu_n = \nu(t_n^-),\ n \geq 1\ .$$

The regeneration instants of the system occur when a newly arriving customer observes an empty system:

$$T_{l+1} = \inf_{n\geq 1}(t_n > T_l : \nu_n = 0), \quad l \geq 0 \ (T_0 := 0) . \tag{4.1.2}$$

4.1.1 Stability Analysis

Denote by

$$\mu_i = 1/\mathsf{E}S^{(i)} , \quad \mu = \sum_{i=1}^{m} \mu_i , \ \rho = \lambda/\mu ,$$

and prove the following stability result.

Theorem 4.1 *Assume that*

$$\rho < 1 , \tag{4.1.3}$$

and

$$\min_{1\leq i\leq m} \min_{1\leq k\leq K} \mathsf{P}(\tau > S_{ik}) > 0 . \tag{4.1.4}$$

Then the process $\{\nu(t)\}$ is positive recurrent for any initial state.

Proof Denote by $A_k(t)$ the number of class-k arrivals in the time interval $[0,\ t]$, and let $A(t) = \sum_{k=1}^{K} A_k(t)$. Also denote by $D_i(t)$ the number of departures from server i in $[0,\ t]$, and let $\widehat{D}_i(t)$ be the number of renewals, during the interval $[0,\ t]$, in the renewal process generated by the service times intended for server i. In order to construct the process $\{\widehat{D}_i(t)\}$, we couple together all service times of server i and then shift them to the origin. Consequently, the process $\{\widehat{D}_i(t)\}$ represents the (zero-delayed) renewal process ($\widehat{D}_i(0) = 1$) corresponding to the case when server i is permanently busy, $i = 1, \ldots, m$. Let $I_i(t)$ be the idle time of server i in $[0,\ t]$,

$$I_i(t) = \int_0^t \mathsf{1}(S_i(u) = 0)du, \ t \geq 0 , \ i = 1, \ldots, m .$$

By construction, $D_i(t) \leq \widehat{D}_i(t)$, and we denote the difference

$$\widehat{D}_i(t) - D_i(t) = \Delta_i(t) , \quad t \geq 0 , \quad i = 1, \ldots, m ,$$

which equals the number of renewals in the process $\{\widehat{D}_i(t)\}$ belonging to the time interval $(t - I_i(t),\ t]$. From the point-of-view of the time-average asymptotic that we consider below, $\Delta_i(t)$ is equivalent to the number of renewals of $\widehat{D}_i(t)$ in the interval $(0,\ I_i(t)]$, (sampled *independently* of $I_i(t)$), and in particular, w.p.1,

$$\lim_{I_i(t)\to\infty} \frac{\Delta_i(t)}{I_i(t)} = \mu_i \,. \tag{4.1.5}$$

Also, denote by

$$D(t) = \sum_{i=1}^{m} D_i(t), \quad \widehat{D}(t) = \sum_{i=1}^{m} \widehat{D}_i(t) \,.$$

Since $A(t) + \nu_1 \geq D(t)$, it then immediately follows from the SLLN that

$$\begin{aligned}\liminf_{t\to\infty} \frac{1}{t}(\widehat{D}(t) - D(t)) &= \liminf_{t\to\infty} \frac{1}{t}\sum_{i=1}^{m} \Delta_i(t) \geq \lim_{t\to\infty} \frac{1}{t}(\widehat{D}(t) - A(t)) \\ &= \mu - \lambda > 0 \,,\end{aligned}$$

implying

$$\liminf_{t\to\infty} \frac{1}{t}\Delta_{i(t)}(t) \geq \frac{\mu - \lambda}{m} > 0 \quad \text{w.p.1} \,, \tag{4.1.6}$$

where index $i(t)$ represents the server with the *largest increment* $\Delta_i(t)$ in the interval $[0,\ t]$. Since the number of renewals in any finite interval is finite w.p.1, it then follows that $I_{i(t)}(t) \to \infty,\ t \to \infty$, w.p.1 (otherwise (4.1.6) would be contradicted), and due to (4.1.5), we obtain

$$\liminf_{t\to\infty} \frac{\Delta_{i(t)}(t)}{I_{i(t)}(t)} \leq \max_i \mu_i < \infty \,.$$

Then

$$\liminf_{t\to\infty} \frac{1}{t}\Delta_{i(t)}(t) = \liminf_{t\to\infty} \frac{\Delta_{i(t)}(t)}{I_{i(t)}(t)} \frac{I_{i(t)}(t)}{t} \leq \max_i \mu_i \liminf_{t\to\infty} \frac{I_{i(t)}(t)}{t} \,, \tag{4.1.7}$$

and then (4.1.6) implies

$$\liminf_{t\to\infty} \frac{1}{t} I_{i(t)}(t) > 0 \,. \tag{4.1.8}$$

Since

$$I(t) := \int_0^t \mathbf{1}(\nu(u) < m)du \geq \max_{1\leq i\leq m} I_i(t) \geq I_{i(t)}(t) \,, \tag{4.1.9}$$

we find from (4.1.8) and (4.1.9) that

$$\liminf_{t\to\infty} \frac{I(t)}{t} > 0 \,, \tag{4.1.10}$$

and, by invoking Fatou's lemma,

$$\liminf_{t\to\infty} \frac{1}{t}\mathsf{E}I(t) > 0 .$$

This, in turn, implies that

$$\inf_n \mathsf{P}(\nu(z_n) < m) \ge \delta_0 , \tag{4.1.11}$$

for some constant $\delta_0 > 0$ and deterministic sequence $z_n \to \infty$. For each server i, the sequence $\{S_i(z_n)\}$ is tight in view of Remark 3.3 (Sect. 3.1). It then follows that, conditioning on the event $\{\nu(z_n) < m\}$, the remaining work sequence,

$$\mathsf{W}(z_n) = \sum_{i=1}^{m} S_i(z_n),\ \ n \ge 1 ,$$

is tight as well implying, together with (4.1.11),

$$\inf_n \mathsf{P}\big(\nu(z_n) < m,\ \mathsf{W}(z_n) \le C\big) \ge \frac{\delta_0}{2} ,$$

for some constant C. Moreover, relying on $\mathsf{E}\tau < \infty$ and condition (4.1.4), there exist positive constants $\varepsilon_0,\ C_0$, such that,

$$\min_{1\le i\le m}\ \min_{1\le k\le K} \mathsf{P}\Big(C_0 \ge \tau > S_{ik} + \varepsilon_0\Big) > 0 . \tag{4.1.12}$$

Then we can continue the proof as in Theorem 2.2 (Sect. 2.2). More precisely, we realize a series of independent events of the form $\{C_0 \ge \tau > S_{ik} + \varepsilon_0\}$, and each one of these events implies that the interarrival time is larger than the service time of the incoming customer by no less than the constant ε_0, and is not larger than a constant C_0. This allows to unload the system and reach an empty state (i.e., a regeneration of the system) in a finite time interval, implying $\mathsf{E}T < \infty$.

Now we prove that $T_1 < \infty$ w.p.1. In view of (4.1.10),

$$I(t) \to \infty,\ \ t \to \infty \ \text{ w.p.1} ,$$

and hence, the total amount of time that the queue-size process $\{\nu(t)\}$ spends in the set $[0,\ m)$ is infinite w.p.1. On the other hand, as it is shown in Theorem 2.3 (Sect. 2.3), this process spends a finite amount of time in any bounded set during a *regeneration period*. (It is easy to check that the proof of Theorem 2.3 remains valid for the system under consideration.) This leads to the conclusion that $T_1 < \infty$ w.p.1, and thus the statement of Theorem 4.1 indeed holds. □

Remark 4.1 Theorem 4.1 can be proved under the assumption

$$\max_{1\le i\le m}\ \max_{1\le k\le K} \mathsf{P}(\tau > S_{ik}) > 0 , \tag{4.1.13}$$

which is much weaker than (4.1.4). In this case the unloading of the system becomes more complicated and is realized via a series of specific class-k_0 customers which are routed to a specific server i_0, satisfying regeneration condition

$$\mathsf{P}(\tau > S_{i_0k_0}) > 0 ,$$

which follows from (4.1.13). (A detailed discussion on this issue can be found in the proof of Theorem 3.2 in [3].)

The proof of the following result is omitted because it is quite similar to the proof for the classical $GI/G/m$ system with identical servers, see Theorems 2.5 and 2.6.

Theorem 4.2 *If* $\lambda > \mu$, *then* $\liminf_{t\to\infty} \nu(t)/t > 0$ *w.p.1 for any initial state. If* $\lambda = \mu$, $\nu_1 = 0$ *and (4.1.4) holds, then* $\nu(t) \Rightarrow \infty$.

We may therefore conclude that the stability conditions formulated in Theorem 4.1 constitute the *stability criteria* of the system with non-identical servers, at least in case of a zero initial state.

4.1.2 Some Special Cases

In this subsection, we verify to which form the stability criteria that were obtained above are reduced for some particular systems.

Consider the original m-server system with K classes of customers, with the additional assumption that the service time of the class-k customers at server i only depends on i, denoted by $S_{ik} = S^{(i)}$, implying that the service times are *class-independent*. In such a case, the system becomes *single-class*, and the service rate at server i is given by $\mu_i = 1/\mathsf{E}S^{(i)}$. In general, the rates μ_i need not to be equal, and thus this setting covers the case of *non-identical* servers. Because of $\sum_i p_i = 1$, it then follows that the stability condition (4.1.3) remains formally the same

$$\lambda < \sum_{i=1}^{m} \mu_i = \mu , \tag{4.1.14}$$

while condition (4.1.4) now becomes $\min_i \mathsf{P}(\tau > S^{(i)}) > 0$.

On the other hand, assume that the service rates only depend on the customer class k, but not on the server index i. This case represents class-dependent service times $\{S_n^{(k)}\}$, with class-k generic time $S^{(k)}$, and we obtain a multiclass system with identical servers. Denoting $\rho_k = \lambda_k \mathsf{E}S^{(k)}$, we further reduce this system to a *single-class* $GI/G/m$ system with generic service time S and

$$\mathsf{E}S = \sum_{k=1}^{K} p_k \mathsf{E}S^{(k)} = \frac{1}{\lambda}\sum_{k=1}^{K} \lambda_k \mathsf{E}S^{(k)} = \frac{1}{\lambda}\sum_{k=1}^{K} \rho_k . \tag{4.1.15}$$

The latter system is positive recurrent if and only if condition $\lambda \mathsf{E} S < m$ holds, which is equivalent to condition

$$\sum_{k=1}^{K} \rho_k < m \,, \tag{4.1.16}$$

with condition (4.1.4) replaced by $\min_k \mathsf{P}(\tau > S^{(k)}) > 0$.

It is straightforward to show that conditions (4.1.16) and $\min_k \mathsf{P}(\tau > S^{(k)}) > 0$ also imply positive recurrence of a K-class m-server system with identical servers and with *work-conserving priorities*, in which the service time of a customer is insensitive to the priority rule [4].

4.2 The Infinite-Server System

In this section, we focus on the *single-class* $GI/G/\infty$ queueing system with an *infinite number of identical* servers, arbitrarily numbered as $1, 2, 3, \ldots$, a renewal input process, and iid service times, see Fig. 4.2. We adopt the same notations as before, with $\nu(t)$ representing the number of customers in the system at time t, which is also equal to the number of busy servers at time t. We also define the set $\mathcal{B}(t) = \{i : S_i(t) > 0\}$ of the numbers of busy servers at time t, so $|\mathcal{B}(t)| = \nu(t)$, and the remaining work satisfies

$$\mathsf{W}(t) = \sum_{i \in \mathcal{B}(t)} S_i(t), \quad t \geq 0 \,.$$

The regeneration instants are defined as in (4.1.2). Also define, assuming it exists, the weak limit $\mathcal{B}(t) \Rightarrow \mathcal{B}$ as the stationary number of busy servers.

At first glance, assumption $\mathsf{P}(\tau > S) > 0$ could seem to be unnecessary in a system with an infinite number of servers. Nonetheless, it is easy to see that without condition $\mathsf{P}(\tau > S) > 0$ classical regenerations (4.1.2) are impossible because the system does not become completely empty.

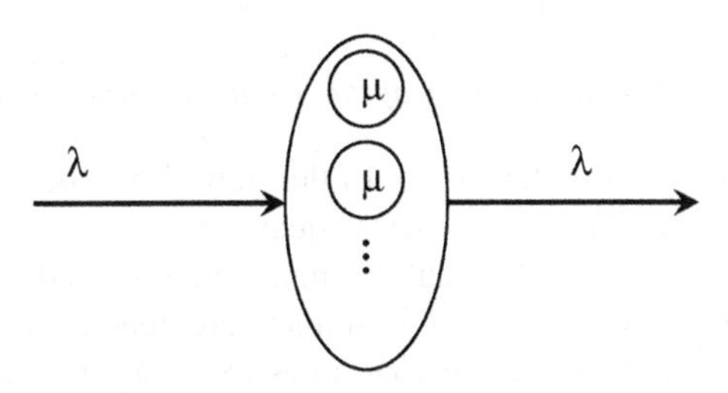

Fig. 4.2 The $GI/G/\infty$ system with an infinite number of servers

Theorem 4.3 *Suppose that*

$$\rho := \lambda \mathsf{E} S < \infty\,, \tag{4.2.1}$$

$$\mathsf{P}(\tau > S) > 0\,. \tag{4.2.2}$$

Then $\mathsf{E}T < \infty$, $T_1 < \infty$ *w.p.1 and*

$$\mathsf{E}\mathcal{B} = \rho. \tag{4.2.3}$$

Proof Denote the original system by Σ, and also consider the auxiliary multiserver *infinite-buffer* $GI/G/\kappa$ queueing system, denoted by $\widetilde{\Sigma}$, constructed as follows. (Whenever appropriate, the corresponding quantities in the auxiliary system are endowed by 'tilde'.) The system $\widetilde{\Sigma}$ is assumed to be fed by the same renewal input, and has the same sequence of service times as the original system Σ, i.e., using coupling,

$$t_n = \widetilde{t}_n, \ \ S_n = \widetilde{S}_n, \ \ n \geq 1\,.$$

Moreover, the number of servers κ in the system $\widetilde{\Sigma}$ is selected in such a way that $\rho < \kappa$, and the positive recurrence of the system $\widetilde{\Sigma}$ follows immediately. Denote by $\widetilde{W}_n^{(1)}$ the waiting time of customer n in the queue of $GI/G/\kappa$ system, and let d_n be the departure instant of customer n in system Σ. Then the ordering

$$d_n = t_n + S_n \leq t_n + \widetilde{W}_n^{(1)} + S_n = \widetilde{d}_n\,,$$

holds, leading to the following inequalities

$$\nu(t) = \#\{k : t_k \leq t < d_k\} \leq \#\{k : t_k \leq t < \widetilde{d}_k\} = \widetilde{\nu}(t), \ \ t \geq 0\,. \tag{4.2.4}$$

Note that $\widetilde{T}_1 < \infty$, $\mathsf{E}\widetilde{T} < \infty$, and that the proof of (4.2.3) is contained in Problem 4.1. Hence, in view of the dominance property (4.2.4), the proof is hereby completed. □

Problem 4.1 *For the positive recurrent system* Σ, *using (1.2.14) and representation*

$$\frac{\mathsf{E}\int_0^T \nu(u)du}{\mathsf{E}T} = \lambda \frac{\mathsf{E}\sum_{k=1}^{\theta} S_k}{\mathsf{E}\theta}\,, \tag{4.2.5}$$

show that the mean stationary number of the occupied servers satisfies (4.2.3).

Note that (4.2.3) is a particular case of Little's law (2.5.4) because, in the original system, the waiting time W_n of customer n equals zero.

The statement of Theorem 4.3 can be immediately extended to the *multiclass infinite-server system* in which class-k customers are generated by a Poisson input process with rate λ_k, and have iid service times $\{S_n^{(k)}\}$. In this case, $\lambda_k \mathsf{E}S^{(k)} = \rho_k$, $\rho = \sum_k \rho_k = \mathsf{E}\mathcal{B}$, and the regeneration condition holds automatically.

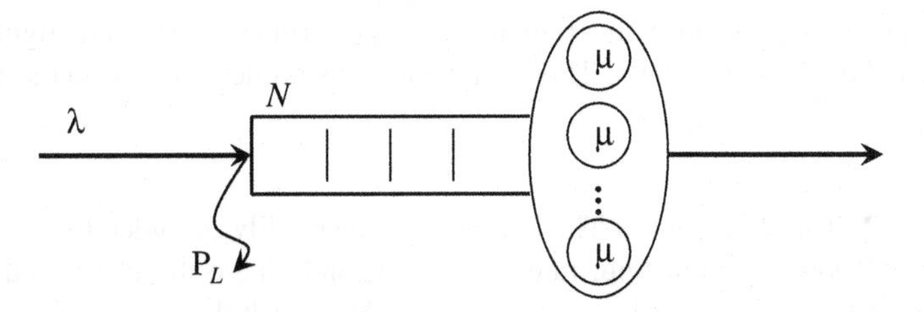

Fig. 4.3 The m-server system $GI/G/m/N$ with finite waiting room N

4.3 The Multiserver Finite-Buffer System

Maintaining our standard notations and assumptions, consider the single-class multiserver $GI/G/m/N$ system with buffer size $N < \infty$ and m *identical servers*. This is a *loss system* in which a customer observing $m + N$ other customers upon arrival is lost (see Fig. 4.3, where P_L represents the stationary *loss probability*). All basic processes describing this system regenerate at the instants (4.1.2).

Because the system size is *finite* (i.e., the number of customers is bounded by $N + m$), it seems tempting to state that this system is always 'stable', even without assumption $\mathsf{P}(\tau > S) > 0$ being satisfied. Indeed, it is straightforward to conclude that the (bufferless) loss system $GI/G/1/0$ regenerates even if $\mathsf{P}(\tau < S) = 1$. The same holds for the 2-server $GI/G/2/0$ system, as the following simple example demonstrates. Let, in a zero initial state system,

$$\mathsf{P}(\tau = 1) = \mathsf{P}(\tau = 2) = \mathsf{P}(S = 4) = \mathsf{P}(S = 5) = 1/2 .$$

Then $\mathsf{P}(\tau < S) = 1$, and as it easy to calculate, customer 5 finds both servers idle with probability

$$\mathsf{P}(\tau_1 = \tau_2 = 1,\ \tau_3 = \tau_4 = 2,\ S_1 = 5,\ S_2 = 4) = 2^{-6} .$$

Hence, in the loss system, the assumption $\mathsf{P}(\tau > S) > 0$ is not necessary to have classical regenerations. Nevertheless, in the following statement we include this assumption which allows to obtain classical regenerations by unloading the system in the same way as we did it in the above analysis.

Theorem 4.4 *Assume that* $\rho = \lambda \mathsf{E}S < \infty$ *and* $\mathsf{P}(\tau > S) > 0$. *Then the* $GI/G/m/N$ *system is positive recurrent regenerative under any initial state.*

Proof The following upper bounds for the queue size and aggregated workload process

$$\nu(t) \le N + m , \qquad \mathsf{W}(t) \le_{st} \sum_{k=1}^{N} S_k + \sum_{i=1}^{m} S_i(t) , \quad t \ge 0 , \tag{4.3.1}$$

hold, respectively. Because the remaining service times $\{S_i(t)\}$ are tight (since $\mathsf{E}S < \infty$), then the statement of Theorem 4.4 follows immediately from the assumption $\mathsf{P}(\tau > S) > 0$. □

Remark 4.2 The statement of Theorem 4.4 can be readily extended to a *multiclass* $GI/G/m/N$ system with input rate $1/\mathsf{E}\tau = \lambda$, and where the class-$k$ customers have input rate $\lambda_k = \lambda p_k$ and iid service times $\{S_n^{(k)}\}$ such that

$$\max_{1\le k\le K} \mathsf{E}S^{(k)} < \infty \quad \text{and} \quad \min_{1\le k\le K} \mathsf{P}(\tau > S^{(k)}) > 0\,.$$

Remark 4.3 In view of (4.3.1), we can deduce that w.p.1

$$\lim_{t\to\infty} \frac{\nu(t)}{t} = \lim_{t\to\infty} \frac{\mathsf{W}(t)}{t} = 0\,.$$

This result is caused by the finiteness of the system. In general, for a (non-negative) tight process $\{X(t)\}$, we may only state the convergence in probability

$$\frac{X(t)}{t} \Rightarrow 0,\ t \to \infty\,. \tag{4.3.2}$$

Indeed, in view of the tightness, for an arbitrary $\varepsilon > 0$, there exists a value x_ε such that, for all $t \ge 0$,

$$\varepsilon \ge \mathsf{P}(X(t) > x_\varepsilon) = \mathsf{P}\Big(\frac{X(t)}{t+1} > \frac{x_\varepsilon}{t+1}\Big)\,,$$

and, for an arbitrary $\delta > 0$, one selects t_δ such that $x_\varepsilon/(t+1) \le \delta$ for $t \ge t_\delta$. Then

$$\sup_{t\ge t_\delta} \mathsf{P}\Big(\frac{X(t)}{t+1} > \delta\Big) \le \varepsilon\,,$$

and then (4.3.2) follows from the arbitrariness of ε and δ. However, it is known, as we also show in (6.2.25) below, that convergence (4.3.2) implies w.p.1 convergence of a *subsequence* $X(z_n)/z_n \to 0$ for a deterministic sequence $z_n \to \infty$.

4.3.1 The Loss Probability

In this subsection, we deduce a relation between the stationary *loss probability* P_L and the stationary *busy probability* P_B of an arbitrary server (in the $GI/G/m/N$ queueing system).

Denote by $R(t)$ the number of lost customers (which observe a fully occupied system upon arrival) during the time interval $[0, t]$, and let $\mathcal{R}(t)$ be the set of indexes of these customers, i.e., $|\mathcal{R}(t)| = R(t)$. Define the indicator function $\mathsf{1}_n = 1$ if customer n is lost, and $\mathsf{1}_n = 0$ otherwise. Due to the aperiodicity of the regeneration period length θ, the stationary loss probability is given by $\mathsf{P}_L = \lim_{n\to\infty} \mathsf{P}(\mathsf{1}_n = 1)$. The number of lost customers

$$R(t) = \sum_{n=1}^{A(t)} \mathsf{1}_n = \sum_{n\in\mathcal{R}(t)} \mathsf{1}_n, \quad t \geq 0\,, \tag{4.3.3}$$

is a positive recurrent process with regenerative increments. Consequently, (1.2.14) leads to the following representation of the loss probability:

$$\mathsf{P}_L = \lim_{t\to\infty} \frac{R(t)/t}{A(t)/t} = \frac{\mathsf{E}\sum_{n=1}^{\theta} \mathsf{1}_n}{\mathsf{E}\theta} = \frac{\mathsf{E}R}{\mathsf{E}\theta}\,, \tag{4.3.4}$$

where R represents the (generic) number of lost customers during a regeneration period. Denoting by $B_i(t)$ the busy time of server i in the interval $[0,\ t]$, we obtain the stationary busy probability as the limit

$$\lim_{t\to\infty} \frac{B_i(t)}{t} = \mathsf{P}_B^{(i)}\,, \quad i = 1, \ldots, m\,.$$

If all servers are identical, then $\mathsf{P}_B \equiv \mathsf{P}_B^{(i)}$ is the stationary busy probability of an arbitrary server.

Theorem 4.5 *In the $GI/G/m/N$ queueing system with identical servers, under conditions $\rho < \infty$, $P(\tau > S) > 0$, the following equality holds:*

$$P_L = 1 - \frac{m}{\rho} P_B\,. \tag{4.3.5}$$

Proof First, in view of Theorem 4.4 the system is positive recurrent. Denote by

$$\mathsf{W}(t) = \sum_{i=1}^{m} W_i(t), \quad t \geq 0\,,$$

the total remaining work at instant t. For the time interval $[0,\ t]$, denote by $V(t)$ the total amount of work that has arrived, and by $L(t)$ the amount of work that has been lost (which equals the total time needed to serve $R(t)$ lost customers). Also define the aggregated busy time $B(t) = \sum_{i=1}^{m} B_i(t)$. Then the following balance equation holds:

$$\mathsf{W}_1 + V(t) = \mathsf{W}(t) + B(t) + L(t)\,,\ t \geq 0\,, \tag{4.3.6}$$

where, by the assumption, $\lim_{t\to\infty} \mathsf{W}(t)/t = 0$ w.p.1. Note that

$$V(t) = \sum_{k=1}^{A(t)} S_k \,, \quad L(t) = \sum_{k \in \mathcal{R}(t)} S_k \,,$$

where the summands $\{S_k\}$ are iid. Then, invoking the SLLN and (4.3.4), we find that, w.p.1,

$$\lim_{t\to\infty} \frac{V(t)}{t} = \rho \,, \tag{4.3.7}$$

and

$$\frac{L(t)}{V(t)} = \frac{\sum_{k\in\mathcal{R}(t)} S_k}{R(t)} \frac{A(t)}{\sum_{k=1}^{A(t)} S_k} \frac{R(t)}{A(t)} \to \frac{\mathsf{E}R}{\mathsf{E}\theta} = \mathsf{P}_L \,, \quad t \to \infty \,.$$

Since the servers are assumed to be identical,

$$\lim_{t\to\infty} \frac{B(t)}{t} = \lim_{t\to\infty} \frac{\sum_{i=1}^{m} B_i(t)}{t} = m\mathsf{P}_B \,.$$

This and (4.3.7) lead to

$$\lim_{t\to\infty} \frac{B(t)}{V(t)} = \frac{m\mathsf{P}_B}{\rho} \,.$$

Consequently, dividing both sides of (4.3.6) by $V(t)$ and letting $t \to \infty$, we indeed arrive at (4.3.5). □

Remark 4.4 It is worth mentioning that, if $\rho = m$, then (4.3.5) becomes

$$\mathsf{P}_L = 1 - \mathsf{P}_B = \mathsf{P}_0 \,, \tag{4.3.8}$$

where P_0 is the *stationary idle probability* of an arbitrary server. This is a well-known result for the $M/M/1/N$ system with $\rho = 1$, in which case the stationary queue size is uniformly distributed and in particular [5],

$$\mathsf{P}_L = \mathsf{P}_0 = \frac{1}{1+N} \,. \tag{4.3.9}$$

Results (4.3.8) and (4.3.9) show that when the mean amount of work ρ that has arrived per unit of time equals the *maximal capacity* (m) of the system (we call this the *driftless* condition), it is equally probable that a newly arriving customer meets an idle server or will be lost.

Relation (4.3.5) can also be established in an alternative, more direct way. Indeed,

$$\mathcal{B}(t) = \sum_{i=1}^{m} 1(S_i(t) > 0) ,$$

is the number of busy servers at instant t, and we set $1_n = 0$ if customer n is accepted, see (4.3.3). It then follows from regenerative arguments that the *mean stationary number of busy servers* is equal to

$$\begin{aligned} \mathsf{E}\mathcal{B} &= \lim_{t\to\infty} \frac{1}{t} \int_0^t \mathcal{B}(u)du = \lambda \frac{\mathsf{E}[\sum_{n=1}^{\theta}(1 - 1_n)S_n]}{\mathsf{E}\theta} \\ &= \rho\Big(1 - \frac{\mathsf{E}\sum_{n=1}^{\theta} 1_n}{\mathsf{E}\theta}\Big) = \rho\,(1 - \mathsf{P}_L) , \end{aligned} \tag{4.3.10}$$

where we have relied on (4.3.4) and the independence between indicator 1_n and service time S_n. The equality (4.3.10), written as

$$\mathsf{E}\mathcal{B} = \lambda(1 - \mathsf{P}_L)\mathsf{E}S , \tag{4.3.11}$$

is a variant of Little's law (2.5.4). For the case of identical servers, $\mathsf{E}\mathcal{B} = m\mathsf{P}_B$, and (4.3.11) transforms to (4.3.5).

Note that the equality (4.3.11) holds for a multiclass system as well in which case the service time S_n has been expressed in terms of the service times of class-k customers as $S_n = \sum_k 1_n^{(k)} S_n^{(k)}$, where $\mathsf{E}1_n^{(k)} = p_k$, see (4.1.15).

4.4 The System with Non-identical Servers

Let us now return to the *single-class* infinite-buffer m-server system introduced in Sect. 4.1.2, assuming *server-dependent* service times. We present an alternative instructive coupling-based proof of stability condition (4.1.14).

In addition to the original system, denoted by Σ, consider a modified system $\widetilde{\Sigma}$, with the same initial state, the same renewal input with rate λ, in which service time $S^{(i)}$ in server i has rate μ_i. Denote by $\mu = \sum_{i=1}^{m} \mu_i$. It is assumed that in system $\widetilde{\Sigma}$, each new arrival joins server i with probability $q_i = \mu_i/\mu$ *regardless of the state of the servers and the system*, $i = 1, \ldots, m$. Recall that $\lambda < \mu$ by assumption. In fact, system $\widetilde{\Sigma}$ corresponds to a collection of single-server systems, where server i has an input rate and traffic intensity given by

$$\lambda_i = \lambda q_i , \quad \rho_i = \frac{\lambda_i}{\mu_i} = \frac{\lambda}{\mu} = \rho < 1, \quad i = 1, \ldots, m , \tag{4.4.1}$$

respectively. (When appropriate, the notation 'tilde' is adopted to designate the corresponding quantity in system $\widetilde{\Sigma}$.) Note that server i of the system $\widetilde{\Sigma}$, operates as a FIFO $GI/G/1$ queueing system with a renewal input, in which the generic interarrival time $\widetilde{\tau}^{(i)}$ can be written as a geometric sum (yielding a geometric convolution for the distribution) of the original interarrival times, see also [6]. More precisely,

$$\widetilde{\tau}^{(i)} =_{st} \sum_{n=1}^{\beta_i} \tau_n ,$$

where the summation bound β_i is independent of the iid summands and has a geometric distribution given by $\mathsf{P}(\beta_i = n) = q_i(1-q_i)^{n-1}$, $n \geq 1$, leading to the mean $\mathsf{E}\widetilde{\tau}^{(i)} = 1/\lambda_i$. In the system Σ, define by $V_i(t)$ the workload that has arrived in the interval $[0, t]$ and is intended for server i, $W_i(t)$ the workload process at instant t in server i, and $I_i(t)$ its idle time during the interval $[0, t]$. Using the coupling approach, we consider input paths coinciding in both systems w.p.1. Then, for each instant t, a server $i(t)$ exists such that the number of customers intended for server $i(t)$ in both systems, among those that arrived in interval $[0,\ t]$, are ordered as

$$A_{i(t)}(t) \leq \widetilde{A}_{i(t)}(t) . \tag{4.4.2}$$

Based on the independent iid sequences $\{S_n^{(i)}\}$, we again use a coupling and assign the *same* service time $S_n^{(i)}$ for the nth customer *entering server i* in both systems $i = 1, \ldots, m$. This immediately implies the following inequality

$$V_{i(t)}(t) = \sum_{k=1}^{A_{i(t)}(t)} S_k^{(i)} \leq \sum_{k=1}^{\widetilde{A}_{i(t)}(t)} S_k^{(i)} = \widetilde{V}_{i(t)}(t), \quad t \geq 0 . \tag{4.4.3}$$

On the other hand, we can write down the following relations:

$$\begin{aligned} W_{i(t)}(0^-) + V_{i(t)}(t) &= t - I_{i(t)}(t) + W_{i(t)}(t) , \\ \widetilde{W}_{i(t)}(0^-) + \widetilde{V}_{i(t)}(t) &= t - \widetilde{I}_{i(t)}(t) + \widetilde{W}_{i(t)}(t) , \end{aligned} \tag{4.4.4}$$

where, by assumption, $W_{i(t)}(0^-) = \widetilde{W}_{i(t)}(0^-)$. Hence, from (4.4.3) and definition of $I(t)$ in (4.1.9), we obtain

$$\widetilde{I}_{i(t)}(t) - \widetilde{W}_{i(t)}(t) \leq I_{i(t)}(t) - W_{i(t)}(t) \leq I_{i(t)}(t) \leq I(t), \quad t \geq 0 .$$

This in turn implies

$$I(t) \geq \min_{1 \leq i \leq m} \widetilde{I}_i(t) - \sum_{i=1}^{m} \widetilde{W}_i(t) , \quad t \geq 0 . \tag{4.4.5}$$

It follows from the construction of system $\widetilde{\Sigma}$ and (4.4.1) that each process $\{\widetilde{W}_i(t)\}$, analyzed separately, is *positive recurrent* regenerative, and in particular (see (2.3.22)),

$$\lim_{t\to\infty} \frac{\widetilde{W}_i(t)}{t} = 0\,. \tag{4.4.6}$$

(Evidently, the regenerations of $\{\widetilde{W}_i(t)\}$ occur when a customer routed to server i finds it idle.) From (4.4.1), the input rate in server i in system $\widetilde{\Sigma}$ satisfies

$$\lim_{t\to\infty} \frac{\widetilde{A}_i(t)}{t} = \lambda q_i = \lambda_i\,.$$

Then, for each $i = 1, \ldots, m$, we obtain

$$\lim_{t\to\infty} \frac{\widetilde{V}_i(t)}{t} = \lim_{t\to\infty} \frac{\widetilde{A}_i(t)}{t} \frac{\sum_{n=1}^{\widetilde{A}_i(t)} S_n^{(i)}}{\widetilde{A}_i(t)} = \rho_i \equiv \rho = \frac{\lambda}{\sum_i \mu_i}\,. \tag{4.4.7}$$

Since the balance relations (4.4.4) hold for each server i (and not only for $i(t)$), then (4.4.7) and (4.4.6) lead to

$$\lim_{t\to\infty} \frac{\widetilde{I}_i(t)}{t} = 1 - \lim_{t\to\infty} \frac{\widetilde{V}_i(t)}{t} = 1 - \rho > 0\,, \quad i = 1, \ldots, m\,,$$

and, due to (4.4.5), we find that

$$\liminf_{t\to\infty} \frac{I(t)}{t} > 0\,.$$

Finally, invoking Fatou's lemma and the discussion following inequality (4.1.11), we once more conclude that assumptions (4.1.14) and $\min_i \mathsf{P}(\tau > S^{(i)}) > 0$ imply the positive recurrence of the system with non-identical servers.

4.5 The Busy Probability and Stationary Remaining Service Time

Next, we derive an expression for the stationary busy probability of a server in the positive recurrent system considered in the previous section.

Consider (for simplicity only) the zero-delayed case, and let us introduce the busy time process of an arbitrary (fixed) server i as

$$B_i(t) = \int_0^t \mathbf{1}(S_i(u) > 0)du, \;\; t \geq 0\,, \;\; i = 1, \ldots, m\,. \tag{4.5.1}$$

Define

$$B_n^{(i)} = B_i(T_n) - B_i(T_{n-1})\,, \; i \geq 1\,,$$

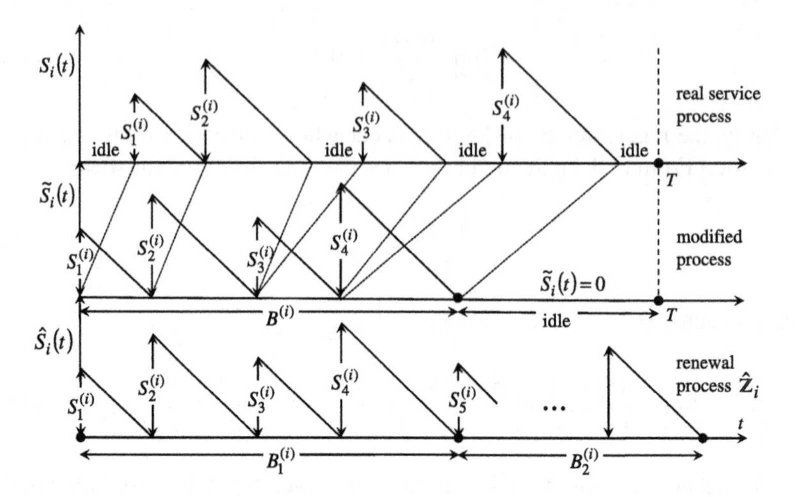

Fig. 4.4 Construction of the modified busy process and renewal process $\widehat{\mathbb{Z}}_i$ using the original service times

the increment of the process (4.5.1) in the nth regeneration cycle of the system, with regeneration instants $\{T_n\}$ defined in (4.1.2). Evidently, the random variables $\{B_n^{(i)}\}$ are iid with a generic increment $B^{(i)}$.

We now modify the system in the following way. First, we 'couple' together all busy periods of server i within each regeneration cycle of the system, and then shift the resulting busy period, distributed as $B^{(i)}$, to the beginning of the corresponding regeneration cycle. This construction process is depicted in Fig. 4.4, and we denote it by

$$\widetilde{B}_i(t) = \int_0^t \mathbf{1}(\widetilde{S}_i(u) > 0)du, \quad t \geq 0,$$

where $\widetilde{S}_i(t)$ is the remaining service time. Also we construct a *renewal process*, denoted by

$$\widehat{\mathbb{Z}}_i = \{\widehat{Z}_n^{(i)} = S_1^{(i)} + \cdots + S_n^{(i)},\ n \geq 1\}, \tag{4.5.2}$$

which consists of the same service times $\{S_n^{(i)}\}$ that appear in the original service process in server i, and let

$$\widehat{S}_i(t) = \min_{n\geq 1}(\widehat{Z}_n^{(i)} - t : \widehat{Z}_n^{(i)} > t), \tag{4.5.3}$$

be the remaining renewal time at instant t. Let F_i represent the distribution function of the service time $S^{(i)}$. Due to the above construction, it is evident that *each service time* $S_n^{(i)}$, as well as *each (composed) busy period* $\widetilde{B}_n^{(i)}$, can be treated as a *regeneration period* of the process $\{\widehat{S}_i(t)\}$. In particular, the renewal points $\{\widehat{Z}_n^{(i)}\}$, are the regeneration instants of the process $\{\widehat{S}_i(t)\}$, see Fig. 4.4. Since both types of regeneration periods have finite means, $\mathsf{E}S^{(i)} < \infty$, $\mathsf{E}B^{(i)} < \infty$, then we obtain from (1.2.14) the following two equivalent representations of the same limit, for any $x \geq 0$,

$$\lim_{t\to\infty} \frac{1}{t}\int_0^t \mathbb{1}(\widehat{S}_i(u) > x)du = \frac{1}{\mathsf{E}B^{(i)}}\mathsf{E}\int_0^{B^{(i)}} \mathbb{1}(\widehat{S}_i(u) > x)du\,, \tag{4.5.4}$$

$$\begin{aligned}\lim_{t\to\infty} \frac{1}{t}\int_0^t \mathbb{1}(\widehat{S}_i(u) > x)du &= \frac{1}{\mathsf{E}S^{(i)}}\mathsf{E}\int_0^{S^{(i)}} \mathbb{1}(\widehat{S}_i(u) > x)du \\ &= \mu_i \int_x^{\infty} (1 - F_i(u))du\,,\end{aligned} \tag{4.5.5}$$

where, in order to obtain the second equality in (4.5.5), we use the equality $\widehat{S}_i(u) = S^{(i)} - u$ if $u < S^{(i)}$. (Also see Problem 2.7 in Sect. 2.5.) Again consider the original service process, and make the following key observation: the stationary busy probability of (each) server i in the original system equals the corresponding quantity in the modified system, because the total busy time of the server within each regeneration period of the system *remains unchanged*, leading to

$$B^{(i)} =_{st} \widetilde{B}^{(i)}\,, \quad i = 1, \ldots, m\,. \tag{4.5.6}$$

Therefore, due to this construction, we find from (1.2.14) that

$$\begin{aligned}&\lim_{t\to\infty} \frac{1}{t}\int_0^t \mathbb{1}(S_i(u) > x)du = \lim_{t\to\infty} \frac{1}{t}\int_0^t \mathbb{1}(\widetilde{S}_i(u) > x)du \\ &= \frac{1}{\mathsf{E}T}\mathsf{E}\int_0^T \mathbb{1}(\widetilde{S}_i(u) > x)du = \frac{1}{\mathsf{E}T}\mathsf{E}\int_0^{B^{(i)}} \mathbb{1}(\widetilde{S}_i(u) > x)du\,,\end{aligned} \tag{4.5.7}$$

where, in order to obtain the last equality, we have taken into account that in the modified system $\widetilde{S}_i(t) = 0$ if $t \in [\widetilde{B}^{(i)}, T)$, see Fig. 4.4. On the other hand, in view of (4.5.6),

$$\mathsf{E}\int_0^{B^{(i)}} \mathbb{1}(\widetilde{S}_i(u) > x)du = \mathsf{E}\int_0^{B^{(i)}} \mathbb{1}(\widehat{S}_i(u) > x)du\,, \tag{4.5.8}$$

because we apply the *same service times* both in the original service process in server i and in the renewal process $\widehat{\mathbb{Z}}_i$. Denote the limit

$$\lim_{t\to\infty} \frac{B_i(t)}{t} = \frac{\mathsf{E}B^{(i)}}{\mathsf{E}T} = \mathsf{P}_B^{(i)}\,, \tag{4.5.9}$$

which is the stationary busy probability of server i,

$$\lim_{t\to\infty} \mathsf{P}(S_i(t) > 0) = \mathsf{P}(\mathbb{S}_i > 0) = \mathsf{P}_B^{(i)}, \tag{4.5.10}$$

when the stationary remaining service time $\mathbb{S}_i$ exists. From expressions (4.5.4)–(4.5.10), we obtain the following statement:

Theorem 4.6 *In the positive recurrent system with non-lattice T,*

$$\begin{aligned} \mathsf{P}(\mathbb{S}_i \le x) &= \lim_{t\to\infty} \frac{1}{t}\int_0^t 1(S_i(u) \le x)du = 1 - \mathsf{P}_B^{(i)}\mu_i \int_x^\infty (1 - F_i(u))du \\ &= \mathsf{P}_0^{(i)} + \mathsf{P}_B^{(i)}\mu_i \int_0^x (1 - F_i(u))du\,, \quad x \ge 0\,, \end{aligned} \tag{4.5.11}$$

where

$$\mathsf{P}_0^{(i)} = 1 - \mathsf{P}_B^{(i)} = \mathsf{P}(\mathbb{S}_i = 0)\,, \ i = 1, \ldots, m\,,$$

represents the stationary idle probability of server i.

In case of identical servers, denoting by $\mathbb{S}_i \equiv \mathbb{S}$, $F_i \equiv F$, $\mu_i \equiv \mu$, the probability $\mathsf{P}_B^{(i)} \equiv \mathsf{P}_B = 1 - \mathsf{P}_0$ is available in an explicit form, see (2.5.29):

$$\mathsf{P}_B = \frac{\rho}{m} = \frac{\lambda}{\mu m}\,, \tag{4.5.12}$$

and then (4.5.11) becomes

$$\mathsf{P}(\mathbb{S} \le x) = 1 - \frac{\lambda}{m} + \frac{\lambda}{m}\int_0^x (1 - F(u))du\,. \tag{4.5.13}$$

Note that since

$$\mathsf{P}(S_i(t) > x) = \mathsf{P}(S_i(t) > x | S_i(t) > 0)\mathsf{P}(S_i(t) > 0)\,,$$

then it follows from (4.5.10) and (4.5.11) that the limit

$$\lim_{t\to\infty} \mathsf{P}(S_i(t) > x | S_i(t) > 0) = \mu_i \int_x^\infty (1 - F_i(u))du\,, \tag{4.5.14}$$

exists and is equal to the *conditional* (tail) distribution of the remaining service time in server i provided that it is busy (see also (2.3.3)).

Next, denote by D_i and I_i respectively the number of departures and the total idle time of server i within a regeneration period of the system. Then

$$\mathsf{P}_0^{(i)} = \frac{\mathsf{E}I_i}{\mathsf{E}T} = 1 - \mathsf{P}_B^{(i)}\,.$$

Since $B_i =_{st} \sum_{k=1}^{D_i} S_k^{(i)}$, then, by Wald's identity, ,

$$\mathsf{E}B^{(i)} = \mathsf{E}\sum_{k=1}^{D_i} S_k^{(i)} = \sum_{k=1}^{\infty} \mathsf{E}[S_k^{(i)} \mathbf{1}(D_i \geq k)] = \mathsf{E}S^{(i)}\mathsf{E}D_i\,, \tag{4.5.15}$$

where we use independence between service time $S_k^{(i)}$ and indicator function $\mathbf{1}(D_i \geq k) = \mathbf{1}(D_i > k-1)$. As a result, we obtain

$$\mathsf{E}T = \mathsf{E}I_i + \mathsf{E}B^{(i)} = \mathsf{E}I_i + \frac{\mathsf{E}D_i}{\mu_i},\ \ 1 \leq i \leq m\,.$$

Therefore,

$$\mu_i(1 - \mathsf{P}_0^{(i)}) = \frac{\mathsf{E}D_i}{\mathsf{E}T}\,,$$

implying the following relations

$$\sum_{i=1}^{m} \mu_i(1 - \mathsf{P}_0^{(i)}) = \sum_{i=1}^{m} \mu_i\,\mathsf{P}_B^{(i)} = \sum_{i=1}^{m} \frac{\mathsf{E}D_i}{\mathsf{E}T} = \lambda\,. \tag{4.5.16}$$

Problem 4.2 *Denote by $D_i(t)$ the number of departures from server i in the interval $[0,\ t]$. Using the balance relation*

$$\nu_1 + A(t) = \sum_{i=1}^{m} D_i(t) + \nu(t),\ \ t \geq 0\,,$$

and a regenerative argument, explain the last equality in (4.5.16).

The relations (4.5.16) also yield the following bounds for $\mathsf{E}\mathcal{B}$, the mean stationary number of busy servers:

$$\frac{\lambda}{\max_i \mu_i} \leq \mathsf{E}\mathcal{B} = \sum_{i=1}^{m} \mathsf{P}_B^{(i)} \leq \frac{\lambda}{\min_i \mu_i}\,. \tag{4.5.17}$$

If the service rates are equal, $\mu_i \equiv \mu$ (but the servers may be *not identical*), then (4.5.17) transforms into $\mathsf{E}\mathcal{B} = \lambda/\mu$.

Remark 4.5 If $\mathsf{E}T = \infty$, which holds for the *non-positive recurrent* system, see Remark 1.1, then $\nu(t) \Rightarrow \infty$, and therefore $\mathsf{P}(S_i(t) > 0) \to 1$ and $\mathsf{E}I_i(t) = o(t)$ for each i. It then follows from the inequalities

$$\int_0^t 1(\widehat{S}_i(u) \le x)du \le \int_0^t 1(S_i(u) \le x)du \le I_i(t) + \int_0^t 1(\widehat{S}_i(u) \le x)du , \qquad (4.5.18)$$

that (see (2.3.3))

$$\lim_{t\to\infty} \frac{1}{t} \int_0^t \mathsf{P}(S_i(u) \le x)du = \mu_i \int_0^x (1 - F_i(u))du , \quad i = 1, \ldots, m , \qquad (4.5.19)$$

where, as before, F_i is the service time distribution in server i. In order to prove (4.5.19), we rely on the w.p.1 convergence of

$$\lim_{t\to\infty} \frac{1}{t} \int_0^t 1(\widehat{S}_i(u) \le x)du = \mu_i \int_0^x (1 - F_i(u))du ,$$

and, due to $\int_0^t 1(\widehat{S}_i(u) \le x)du/t \le 1$, also use the dominated convergence theorem and finally apply the convergence in mean in the inequalities (4.5.18). Another approach, based on the Borel-Cantelli lemma, is to consider a deterministic subsequence $z_n \to \infty$ such that $I_i(z_n)/z_n \to 0$ w.p.1, and then take the limit w.p.1 in both inequalities in (4.5.18) along the sequence $\{z_n\}$. (For more detail, also see the discussion preceding (6.2.25).) Consequently, in this case the time-average limit of the remaining service time distribution in server i coincides with the distribution of the *stationary remaining renewal time* in the renewal process (4.5.2). This result has a clear intuitive interpretation.

4.5.1 Extension to a Multiclass System

In this section, we extend Theorem 4.6 to the FCFS *multiclass* multiserver $M/G/m$ system with m *identical servers* and K classes of customers, in which class-k customers are generated by a Poisson input process with rate λ_k, $k = 1, \ldots, K$. We denote by $\lambda = \sum_k \lambda_k$ the rate of the superposed (Poisson) input with the arrival instants $\{t_n\}$. Evidently, this system regenerates when a new arrival observes it completely idle. It is assumed that the service times $\{S_n^{(k)}\}$ of class-k customers are iid with generic element $S^{(k)}$, service rate $\mu_k = 1/\mathsf{E}S^{(k)}$ and distribution function G_k. Let $\nu(t)$ be the total number of customers in the system at instant t, and denote by $S(t)$ the remaining service time at instant t. Because servers are identical, then a few following limiting results will not be dependent on the server index, and we will omit the server index in our notation. Also denote by $S_k(t)$ the remaining service time at instant t of a class-k customer in service. By definition, we set $S_k(t) = 0$ if, at instant

t, the server is either *empty or serves a class-j customer*, $j \neq k$. Then the following equality holds

$$\mathbb{1}(S(t) > x) = \sum_{k=1}^{K} \mathbb{1}(S_k(t) > x)\ ,\ x \geq 0\ ,\ t \geq 0\ , \tag{4.5.20}$$

in which, at each instant t, at most one summand is positive (and equals 1). Because

$$\mathsf{P}(S_k(t) > x) = \mathsf{P}(S_k(t) > x | S_k(t) > 0)\mathsf{P}(S_k(t) > 0)\ , \tag{4.5.21}$$

then equality (4.5.20) implies

$$\mathsf{P}(S(t) > x) = \sum_{k=1}^{K} \mathsf{P}(S_k(t) > x | S_k(t) > 0)\mathsf{P}(S_k(t) > 0)\ . \tag{4.5.22}$$

Now we assume that the system is positive recurrent, then the limit

$$S_k(t) \Rightarrow \mathbb{S}_k\ ,\ t \to \infty\ ,\ k = 1, \ldots, K\ ,$$

exists and is the stationary remaining service time of a class-k customer. Also let $S(t) \Rightarrow \mathbb{S}$ be the stationary unconditional remaining service time (in an arbitrary server), and denote by

$$\rho_k = \lambda_k \mathsf{E}S^{(k)},\ \ \rho = \sum_{k=1}^{K} \rho_k\ .$$

Now we prove the following generalization of Theorem 4.6. Denote by $\mathsf{P}_b^{(k)}$ the stationary probability that an arbitrary server is occupied by a class-k customer, and let $\mathsf{P}_0 = 1 - \sum_{k=1}^{K} \mathsf{P}_b^{(k)}$ be the stationary idle probability of an arbitrary server. (The notation $\mathsf{P}_b^{(k)}$ is adopted here to distinguish from the busy probability $\mathsf{P}_B^{(i)}$ of server i.)

Theorem 4.7 *If $\rho < m$, then the system is positive recurrent for any initial state, and*

$$P(\mathbb{S} \leq x) = P_0 + \sum_{k=1}^{K} P_b^{(k)} \mu_k \int_0^x (1 - G_k(u))du\ , \tag{4.5.23}$$

where $P_b^{(k)} = \rho_k/m$, $k = 1, \ldots, K$.

Proof To prove the positive recurrence, we note that a new customer arriving in the system is of class-k with probability $p := \lambda_k/\lambda$. Then the system becomes single-class with mean service time $\mathsf{E}S = \sum_{k=1}^{K} p_k \mathsf{E}S^{(k)}$, and assumption $\rho < m$ becomes

$$\lambda \mathsf{E}S = \lambda \sum_{i=1}^{K} p_k \mathsf{E}S^{(k)} = \sum_{i=1}^{K} \rho_k < m ,$$

implying the positive recurrence of the system. (In this system the regeneration condition holds automatically.) Introduce the busy time process

$$B_k(t) = \int_0^t \mathbf{1}(S_k(u) > 0)du , \quad t \geq 0 . \tag{4.5.24}$$

Then

$$\begin{aligned} \mathsf{P}_b^{(k)} := \lim_{t\to\infty} \frac{B_k(t)}{t} &= \lim_{t\to\infty} \mathsf{P}(S_k(t) > 0) \\ &= \mathsf{P}(\mathbb{S}_k > 0) , \quad k = 1, \ldots, K , \end{aligned} \tag{4.5.25}$$

and, as in the proof of Theorem 4.6,

$$\begin{aligned} \lim_{t\to\infty} \mathsf{P}(S_k(t) > x) &= \lim_{t\to\infty} \frac{1}{t} \int_0^t \mathbf{1}(S_k(u) > x)du \\ &= \mathsf{P}_b^{(k)} \mu_k \int_x^\infty (1 - G_k(u))du . \end{aligned} \tag{4.5.26}$$

In view of (4.5.11), (4.5.22), it implies (4.5.23).

The number of class-k arrivals in the interval $[0,\ t]$ satisfies $A_k(t)/t \to \lambda_k$, and each server handles a fraction λ_k/m of class-k customers. The total amount of work $V_k(t)$ that class-k customers bring to the system during $[0,\ t]$ satisfies the equation

$$W_k(0^-) + V_k(t) \equiv W_k(0^-) + \sum_{n=1}^{A_k(t)} S_n^{(k)} = W_k(t) + \sum_{i=1}^{m} B_k^{(i)}(t) , \tag{4.5.27}$$

where $W_k(t)$ is the remaining *class-k amount of work* at instant t, and $B_k^{(i)}(t)$ is the total time, in the interval $[0,\ t]$, during which *server* i is occupied by class-k customers, $k = 1, \ldots, K$. (Unlike (4.5.24) and (4.5.25), we now need the additional index to refer to the busy time of server i.) It immediately follows from (4.5.27) that

$$\lambda_k \mathsf{E}S^{(k)} = \rho_k = m\mathsf{P}_b^{(k)}, \tag{4.5.28}$$

where, in order to obtain the 2nd equality, we take into account that the sum $\sum_{i=1}^m B_k^{(i)}(t)$ is *equally* divided (in the limit) among m identical servers, and the proof is hereby completed. □

Because of the relation (4.5.21), it then follows from (4.5.25), (4.5.26) that the limit

$$\lim_{t\to\infty} \mathsf{P}(S_k(t) > x | S_k(t) > 0) = \mu_k \int_x^\infty (1 - G_k(u))du ,$$

is the tail distribution of the remaining service time provided that the server is occupied by a class-k customer.

Problem 4.3 *Using the equality (and explain!)*

$$\mu_k \int_0^\infty (1 - G_k(u))du = 1 ,$$

and also (4.5.28), show that

$$\mathsf{P}(\mathbb{S} \leq x) = \mathsf{P}_0 + \sum_{k=1}^K \frac{\lambda_k}{m} \int_0^x (1 - G_k(u))du , \quad x \geq 0 . \tag{4.5.29}$$

(See also (4.5.11) and (4.5.13) in this regard.)

4.5.2 Non-positive Recurrent System

Assume now that the system is *non-positive recurrent*, $\mathsf{E}T = \infty$, in which case

$$\lim_{t\to\infty} \mathsf{P}(\nu(t) > k) = 1 \ \text{ for each } k \geq 0 ,$$

implying $\mathsf{P}(S(t) > 0) \to 1$, and condition $\rho = \sum_{k=1}^K \rho_k \geq m$ must hold. Then in the subsequent asymptotic analysis we can ignore the idle time of each server, see (4.5.18), (4.5.19). Because the service discipline is FCFS, then an arbitrary (new) customer assigned to a given server is of class-k with probability $p_k = \lambda_k/\lambda$. Consequently, the limiting fraction of the *class-k work* assigned for each server equals the limiting fraction of the total class-k work $V_k(t)$ among all the work $V(t)$ received in $[0, t]$, i.e., each server devotes to class-k customers the following fraction of busy time:

$$\lim_{t\to\infty} \frac{V_k(t)}{V(t)} = \frac{\rho_k}{\rho} =: \widehat{\mathsf{P}}_b^{(k)} ,$$

which can therefore be interpreted as the 'limiting probability' that an arbitrary server is occupied by a class-k customer. (Note that if all customers have the same mean service time, $\mu_k \equiv \mu$, then $\widehat{\mathsf{P}}_b^{(k)} = p_k$.) Consequently, in this case we must replace in (4.5.26) the probability $\mathsf{P}_b^{(k)}$ by $\widehat{\mathsf{P}}_b^{(k)}$ to obtain, by analogy with (4.5.18) and (4.5.19), the following time-average limit of the remaining service time (tail) distribution of a class-k customer:

$$\lim_{t\to\infty}\frac{1}{t}\int_0^t \mathsf{P}(S_k(u) > x)du$$
$$= \frac{\lambda_k}{\rho}\int_x^\infty (1 - G_k(u))du, \; k = 1, \dots, K\,, \tag{4.5.30}$$

and therefore the time-average limit of the unconditional service time distribution is given by

$$\lim_{t\to\infty}\frac{1}{t}\int_0^t \mathsf{P}(S(u) \le x)du = 1 - \sum_{k=1}^{K}\frac{\lambda_k}{\rho}\int_x^\infty (1 - G_k(u))du$$
$$= \sum_{k=1}^{K}\frac{\rho_k}{\rho}\mu_k\int_0^x (1 - G_k(u))du\,, \; x \ge 0\,, \tag{4.5.31}$$

where we take into account that $\rho = \sum_k \rho_k$.

Remark 4.6 In the single-server single-class $M/G/1$ system, with iid service times with distribution function F and service rate $\mu = 1/\mathsf{E}S$, expression (4.5.31) becomes

$$\lim_{t\to\infty}\frac{1}{t}\int_0^t \mathsf{P}(S(u) \le x)du = 1 - \mu\int_x^\infty (1 - F(u))du = \mu\int_0^x (1 - F(u))du,\; x \ge 0\,,$$

and coincides with the distribution (2.3.3) of the stationary remaining renewal time in the process generated by the service times. Note that we can not claim the weak convergence $S_k(t) \Rightarrow \mathbb{S}_k,\; S(t) \Rightarrow \mathbb{S}$ in the non-positive recurrent case. Also, as we observe from (4.5.31), in this case the number of servers m does no longer play a role in the limiting distribution of the remaining service time.

4.6 Regenerative Input Processes

Consider an m-server queueing system with identical servers, iid service times $\{S_n\}$ and interarrival times $\tau_n = t_{n+1} - t_n,\; n \ge 1$. Here we assume that the input process is (zero-delayed) *regenerative*, i.e., there exists an increasing sequence of integer-valued random variables

$$1 < \alpha_1 < \alpha_2 < \cdots\,,$$

such that the input sequence can be divided into iid groups,

$$(\tau_1, \dots, \tau_{\alpha_1 - 1}),\; (\tau_{\alpha_1}, \dots, \tau_{\alpha_2 - 1}), \dots, (\tau_{\alpha_n}, \dots, \tau_{\alpha_{n+1}-1}), \dots\;,$$

called regeneration cycles, with iid cycle lengths $\alpha_{n+1} - \alpha_n$, and with a generic length α. If $\alpha_n \equiv 1$, then the input process reduces to a renewal process. When considering continuous-time processes related to the queueing behaviour, the regenerations of the input process are defined as $\gamma_n = t_{\alpha_n}$ with iid lengths of regeneration cycles $\gamma_{n+1} - \gamma_n$, $n \geq 1$, and a generic length γ. We refer to the customers that start a new regeneration cycle of the input process as *regenerative*. The regenerative input is called *positive recurrent* if $\mathsf{E}\alpha < \infty$, $\mathsf{E}\gamma < \infty$. (An extension to *delayed* regenerative input is obvious [1].) Denote by ν_k the number of customers in the system at arrival instant t_k^- and by $A(t)$ the number of arrivals in $[0,\ t]$, then, in particular, $\nu_{\alpha_n} = \nu(t_{\alpha_n}^-) = \nu(\gamma_n^-)$. Because $\{A(t)\}$ is a non-decreasing process with regenerative increments, it then follows that, for the positive recurrent case, the input rate exists and equals

$$\lim_{t\to\infty} \frac{A(t)}{t} = \frac{\mathsf{E}\alpha}{\mathsf{E}\gamma} =: \Lambda \in (0,\ \infty)\ .$$

Then the regeneration instants of the corresponding discrete- and continuous-time queueing processes are given by

$$\begin{aligned} \theta_{n+1} &= \min\{\alpha_k > \theta_n\ :\ \nu_{\alpha_k} = 0\}\ ,\quad n \geq 0\ ,\quad \theta_0 := 1\ , \\ T_{n+1} &= \min_k\{\gamma_k > T_n\ :\ \nu_{\alpha_k} = 0\}\ ,\quad n \geq 0\ ,\quad T_0 := t_1 = 0\ , \end{aligned}$$

with generic periods θ and T, respectively. Hence, a regeneration is initiated when a *regenerative customer arrives in an empty system*. The detailed proof of the following statement can be found in [1].

Theorem 4.8 *If the regenerative input is positive recurrent,* $\Lambda\mathsf{E}S < m$ *and*

$$\mathsf{P}(\tau_1 > S,\ \alpha = 1) > 0\ , \tag{4.6.1}$$

then $\mathsf{E}T < \infty$ *and* $\mathsf{E}\theta < \infty$.

An important step of the proof is that, due to the event $\{\alpha = 1\}$ in the regeneration condition (4.6.1), we can realize the input process in such a way that, after the process hits a bounded set, each customer in an arbitrary long series of arrivals is *regenerative*. In other words, in this series of arrivals, each regeneration cycle of the input contains exactly one customer and the input process becomes a *renewal* process. This allows in a standard way, using the event $\{\tau_1 > S\}$ in the assumption (4.6.1), to unload the system and start a new regeneration cycle of the system in a limited time with a positive probability.

4.7 Notes

Relation (4.2.5) for the system with stationary input can be found in [7] as well. It seems tempting to apply (4.3.5) to estimate the unknown probability P_L via an estimate of the idle probability P_B when the former, being extremely small, is hard to estimate reliably by means of simulation; this is the case, for instance, if the buffer is very large. (Indeed, this is a typical problem that is addressed by *rare event simulation*, see for instance [8].) However, as it is shown in [9, 10], the variance of the sample mean estimate of P_L increases as the buffer increases, making the estimate inaccurate. In this respect, among the most relevant papers, see [1].

We mention a modification of the basic multiclass system, in which an interchanging of positions of some of the waiting customers is allowed, but where the limits $A_k(t)/t \to p_k$ remain unchanged for all k. In this case, the stability condition (4.1.2) still holds. The latter system relates to the so-called *discriminatory service discipline*, containing a trade-off between queues of different classes (for $K = 2$ classes, see [11]). Such a system is also closely related to the system with *relative priority* (see [12] and references therein). Another application is the system with *multiple reservations* [13, 14].

It is shown in [15], where a two-server system with server- and class-dependent service times is considered, that for the case of more than 2 servers, the stability condition may depend on the entire service time distribution functions, and not merely on the first moment(s) (also see [16]).

As we mentioned in Sect. 2.5, elimination of condition (4.2.2) still allows us to prove both the tightness and positive recurrence but only in terms of the one-dependent regeneration [2, 17]. (Also see Sect. 10.4.)

The material presented in Sect. 4.5 is mainly based on results that can be found in [18]. It is possible to extend the results of Sect. 4.5.1 to the case of non-reliable servers, by using the notion of *generalized service time*, which include all *setup times* caused by the interruptions that happened during the service of a customer; for more detail see [19]. The distribution (4.5.29) for the single-server single-class system with Poisson input is in [4] (formula (10)).

An important application of the system with regenerative input arises from the observation that such an input crossing a queueing network with no *overloaded stations*, preserves the regenerative property, see for instance, papers [1, 20–23] (which mainly consider *one-dependent regenerations*). Therefore, a regenerative input process is a convenient and general construction that enables the description of a wide class of customer streams in such networks. We briefly discuss regenerative networks in Chap. 10. The papers [24, 25] consider a different concept of a regenerative input process (summarized in [26]), where a regeneration point not necessary coincides with an arrival instant.

References

1. Morozov, E.: Weak regeneration in modeling of queueing processes. Queueing Syst. **46**, 295–315 (2004)
2. Sigman, K.: One-dependent regenerative processes and queues in continuous time. Math. Oper. Res. **15**, 175–189 (1990)
3. Morozov E (2002) Instability of nonhomogeneous queueing networks
4. Wolff, R.W.: Work-conserving priorities. J. Appl. Prob. **7**(2), 327–337 (1970)

5. Asmussen, S.: Applied Probability and Queues, 2nd edn. Springer, New York (2003)
6. Kalashnikov, V.: Topics on Regenerative Processes. CRC Press, Roca Baton (1994)
7. Baccelli, F., Bremaud, P.: Elements of Queueing Theory: Palm Martingale Calculus and Stochastic Recurrences, 2nd edn. Springer, New York (2003)
8. Heidelberger, P.: Fast simulation of rare events in queuieng and relaibility models. Performance Evaluation of Computers and Communications Systems, LN in Computer Science, vol. 729, pp. 165–202. Springer, Berlin (1993)
9. Srikant, R., Whitt, W.: Variance reduction in simulations of loss models. Oper. Res. **47**(4), 509–523 (1999)
10. Whitt, W.: A review of $L = \lambda W$ and extensions. Queueing Syst. **9**, 235–268 (1991)
11. Kim, J., Kim, B.: A survey of retrial queueing systems. Ann. Oper. Res. **247**(1), 3–36 (2016)
12. Ayesta, U., Izagirre, A., Verloop, I.M.: Heavy-traffic analysis of the discriminatory random-order-of-service discipline. ACM Sig. Perf. Eval. Rev. **39**(2), 41–43 (2011)
13. De Vuyst, S., Wittevrongel, S., Bruneel, H.: Place reservation: delay analysis of a novel scheduling mechanism. Comp. Oper. Res. **35**(8), 2447–2462 (2008)
14. Feyaerts, B., De Vuyst, S., Bruneel, H., Wittevrongel, S.: Delay analysis of a discrete-time GI/GI/1 queue with reservation-based priority scheduling. Stoch. Mod. **32**(2), 179–205 (2016)
15. Foss, S., Chernova, N.: On the stability of a partially accessible multi-station queue with state-dependent routing. Queueing Syst. **29**, 55–73 (1998)
16. Dai, J., Hasenbein, J., Kim, B.: Stability of join-the-shortest-queue networks. Queueing Syst. **57**, 129–145 (2007)
17. Sigman, K., Wolff, R.W.: A review of regenerative processes. SIAM Rev. 35(2), 269–288 (1993)
18. Morozov, E., Morozova, T.: On the stationary remaining service time in the queueing systems. In: Proceedings Second International Workshop on Stochastic Modeling and Applied Research of Technology (SMARTY 2020, CEUR-WS 2792), pp. 140–149, Petrozavodsk (2020)
19. Dimitriou, I., Morozov, E., Morozova, T.: A multiclass retrial system with coupled orbits and service interruptions: verification of stability conditions. In: Proceedings of the $24th$ Conference of Open Innovations Association, FRUCT, 2019 ISSN: 2305-7254 eISSN: 2343-0737, vol. 24, pp. 75–81 (2019)
20. Kalashnikov, V., Rachev, S.: Mathematical Methods for Construction of Queueing Models. The Wadsworth and Brooks/Cole Mathematical Series. Springer, New York (1990)
21. Nummelin, E.: A conservation property for general $GI|G|1$ queues with application to tandem queues. Adv. Appl. Prob. **11**, 660–672 (1979)
22. Nummelin, E.: Regeneration in tandem queues. Adv. Appl. Prob. **13**, 221–230 (1981)
23. Sigman, K.: The stability of open queueing networks. Stoch. Proc. Appl. **35**, 11–25 (1990)
24. Afanasyeva, L.G., Bashtova, E.E.: Coupling method for asymptotic analysis of queues with regenerative input and unreliable server. Queueing Syst. **76**, 125–147 (2014)
25. Afanasyeva, L.G.: Asymptotic analysis of queueing models based on synchronization method. Method. Comp. Appl. Prob. **22**, 1417–1438 (2020)
26. Afanasyeva, L.G.: Stability analysis of queueing systems based on synchronization of the input and majorizing output flows. In: Anisimov, V., Limnios, N. (eds.) Queueing Theory 2, Advanced Trends, 1–32. Wiley/ISTE, New York (2021)

Chapter 5
State-Dependent Systems

In this chapter, we apply the regenerative stability analysis to a number of queueing systems that belong to a class of *state-dependent* systems. In each of these systems, a mechanism is available that allows to control its dynamics, depending on the current state of the basic queueing processes describing the system.

We consider a few basic models, highlighting the associated new methodological features of their stability analysis. Indeed, the analysis of these state-dependent systems is possibly the best illustration of an important feature of the regenerative stability method: in order to establish positive recurrence, it is sufficient to prove the negative drift of a basic process *outside a bounded set* (containing a *regeneration state*), that, as a rule, implies that the system approaches a saturated regime. (This negative drift concept is a basic element of the stability analysis of Markov chains, see [1].)

5.1 A Workload-Dependent System

In this section, we will consider a single-server state-dependent system with a dependency between the workload process on the one hand, and the service time and interarrival time process on the other hand.

Consider the following FIFO single-server $G/G/1$ system, in which the interarrival time $\tau_n = t_{n+1} - t_n$ between the nth and $(n+1)$th customers, and the service time S_n of customer n, depend on the workload W_n observed by customer n upon arrival, $n \geq 1$. More precisely, assume that if $W_n = x$, then S_n and τ_n are distributed as random variables $S(x)$ and $\tau(x)$, respectively, and their predefined (conditional) distribution functions depend on x only. In other words, for a given value of x, $S(x)$ and $\tau(x)$ are assumed to be mutually independent random variables. Evidently, such a system covers a wide class of systems where a control mechanism for the input and/or service processes is provided, governed by the current remaining work. It

E. Morozov and B. Steyaert, *Stability Analysis of Regenerative Queueing Models*,
https://doi.org/10.1007/978-3-030-82438-9_5

stands to reason that such a control mechanism should ensure that the higher the actual remaining work becomes, the less new work should be allowed to enter and/or the higher the service effort becomes. Define

$$U(x) = S(x) - \tau(x),$$

and note that the embedded workload process sequence $\{W_n\}$ is still a Markov chain satisfying the following modified *Lindley's recursion*

$$W_{n+1} = \big(W_n + U(W_n)\big)^+, \quad n \geqslant 1.$$

It is now easy to deduce, due to the mechanism determining the dependency, that

$$\theta_{n+1} = \min_{k\geq 1}\{\, k > \theta_n : W_k = 0 \,\}, \quad n \geq 0, \quad \theta_0 := 1, \tag{5.1.1}$$

are still the regeneration instants of the sequence $\{W_n\}$, with generic regeneration period θ. (Regenerations in a continuous-time setting can be defined as before as well, however we will not use this construction in what follows.) The following statement establishes the sufficient positive recurrence conditions of the zero initial state system:

Theorem 5.1 *Assume that the set of random variables $\{\tau(x),\ x \geqslant 0\}$ is uniformly integrable. Moreover, assume that the following assumptions hold:*

$$\sup_{x\geqslant 0} \mathsf{E} S(x) < \infty; \tag{5.1.2}$$

$$\limsup_{x\to\infty} \mathsf{E} U(x) =: -\varepsilon_0 < 0; \tag{5.1.3}$$

for any constant $D \geqslant 0$, there exist constants $\varepsilon > 0,\ \delta > 0$ such that

$$\inf_{x\leq D} \mathsf{P}(\tau(x) > S(x) + \delta) \geqslant \varepsilon. \tag{5.1.4}$$

Then $\mathsf{E}\theta < \infty$.

Proof It is easy to recognize (5.1.3) as a negative drift condition, while (5.1.4) is a regeneration condition. Evidently,

$$-\tau(x) \leq U(x) \leq S(x), \ x \geq 0. \tag{5.1.5}$$

Denote by $\Delta_n = W_{n+1} - W_n$, then for any $x \geq 0$,

$$\mathsf{E}\Delta_n = \mathsf{E}(\Delta_n;\ W_n \leq x) + \mathsf{E}(\Delta_n;\ W_n > x), \tag{5.1.6}$$

while the value of Δ_n conditioned on the event $\{W_n = x\}$ and denoted by $\Delta(x)$ has the conditional expectation

$$\begin{aligned}\mathsf{E}\Delta(x) = \mathsf{E}(\Delta_n \mid W_n = x) &= \mathsf{E}\Big((x + U(x))^+ - x\Big) \\ &= -x\ \mathsf{P}(U(x) \le -x) + \mathsf{E}(U(x);\ U(x) > -x)\,.\end{aligned} \tag{5.1.7}$$

From (5.1.2), (5.1.5) and (5.1.7), we obtain

$$\begin{aligned}\sup_{x\ge 0}\ \mathsf{E}\Delta(x) &\le \sup_{x\ge 0}\ \mathsf{E}(U(x);\ U(x) > -x) \\ &\le \sup_{x\ge 0}\ \mathsf{E}[S(x)\ 1(U(x) > -x)] \le \sup_{x\ge 0}\ \mathsf{E}S(x) =: C_0 < \infty\,.\end{aligned} \tag{5.1.8}$$

Recall that the uniform integrability of $\{\tau(x)\}$ means that, see [2],

$$\lim_{y\to\infty}\ \sup_{x\ge 0}\ \mathsf{E}(\tau(x);\ \tau(x) > y) = 0\,.$$

Then, for each $\delta^* > 0$, there exists a value x^* such that

$$\begin{aligned}\delta^* &\ge \sup_{u\ge 0}\ \mathsf{E}(\tau(u);\ \tau(u) > x^*) = \sup_{u\ge 0}\int_{y>x^*} y\ \mathsf{P}(\tau(u)\in dy) \\ &\ge \sup_{u\ge x^*}\int_{y>x^*} y\ \mathsf{P}(\tau(u)\in dy) \ge x^*\ \sup_{u\ge x^*}\ \mathsf{P}(\tau(u) > x^*) \ge x^*\ \mathsf{P}(\tau(x^*) > x^*)\,.\end{aligned}$$

Since δ^* is arbitrarily chosen, it follows that

$$\lim_{x\to\infty} x\ \mathsf{P}(\tau(x) > x) = 0\,, \tag{5.1.9}$$

and hence, $\lim_{x\to\infty} x\ \mathsf{P}\big(\tau(x) > S(x) + x\big) = 0$. Consequently, as $x \to \infty$,

$$-\,x\ \mathsf{P}(U(x) \le -x) = -x\ \mathsf{P}\Big(\tau(x) \ge S(x) + x\Big) \to 0\,. \tag{5.1.10}$$

On the other hand,

$$\mathsf{E}(U(x);\ U(x) > -x) = \mathsf{E}U(x) - \mathsf{E}\big(U(x);\, U(x) \le -x\big)\,. \tag{5.1.11}$$

Relying on (5.1.8) and (5.1.9), we obtain, as $x \to \infty$,

$$\begin{aligned}\mathsf{E}(U(x); U(x) \le -x) &\le \mathsf{E}\big(S(x); \tau(x) > x + S(x)\big) \le \mathsf{E}\big(S(x); \tau(x) > x\big) \\ &\le \mathsf{E}S(x)\ \mathsf{P}(\tau(x) > x) \le C_0\ \mathsf{P}(\tau(x) > x) \to 0\,,\end{aligned} \tag{5.1.12}$$

where, in order to obtain the 3rd inequality, the independence between $\tau(x)$ and $S(x)$ was taken into account. It then follows from (5.1.3), (5.1.7) and (5.1.10)–(5.1.12) that

$$\limsup_{x\to\infty} \mathsf{E}\Delta(x) = \limsup_{x\to\infty} \mathsf{E}(U(x);\ U(x) > -x) = \limsup_{x\to\infty} \mathsf{E}U(x) = -\varepsilon_0\,. \tag{5.1.13}$$

Therefore, there exists a value x_0 such that

$$\mathsf{E}\Delta(x) \le -\varepsilon_0/2 \ \text{ for } \ x \ge x_0\,,$$

implying

$$\mathsf{E}(\Delta_n;\ W_n > x_0) = \int_{x_0}^{\infty} \mathsf{E}\Delta(y)\ \mathsf{P}(W_n \in dy) \le -\frac{\varepsilon_0}{2}\ \mathsf{P}(W_n > x_0)\,. \tag{5.1.14}$$

On the other hand,

$$\begin{aligned}\mathsf{E}(\Delta_n;\ W_n \le x_0) &= \int_0^{x_0} \mathsf{E}\Delta(y)\ \mathsf{P}(W_n \in dy) \le \int_0^{x_0} \mathsf{E}S(y)\ \mathsf{P}(W_n \in dy) \\ &\le C_0\ \mathsf{P}(W_n \le x_0)\,.\end{aligned}$$

This result combined with (5.1.6) and (5.1.14) leads to

$$\mathsf{E}\Delta_n \le C_0\ \mathsf{P}(W_n \le x_0) - \frac{\varepsilon_0}{2}\ \mathsf{P}(W_n > x_0)\,. \tag{5.1.15}$$

Next, assume for a moment that

$$W_n \Rightarrow \infty\,,\ n \to \infty\,. \tag{5.1.16}$$

Since this would imply

$$C_0\ \mathsf{P}(W_n \le x_0) \to 0 \ \text{ and } \ \mathsf{P}(W_n > x_0) \to 1\,,$$

it then follows from (5.1.15) that

$$\limsup_{n\to\infty} \mathsf{E}\Delta_n < 0\,. \tag{5.1.17}$$

This contradicts (5.1.16), and therefore there exist constants D_0, $\delta_0 > 0$, and a deterministic sequence $n_i \to \infty$ such that

$$\inf_i \mathsf{P}(W_{n_i} \le D_0) \geqslant \delta_0 . \tag{5.1.18}$$

Define the integer $L = \lceil D_0/\delta \rceil$, where $\delta > 0$ satisfies assumption (5.1.4) with $D = D_0$. As a next step, fix some value n_i, and consider the event

$$\{W_{n_i} \le D_0\} \bigcap \bigcap_{k=0}^{L-1} \{\tau_{n_i+k} > S_{n_i+k} + \delta\} =: \mathcal{A}_i(L) . \tag{5.1.19}$$

Conditioning on this event, the remaining work (as long as it is strictly positive) decreases during each interarrival time by no less than the value δ, implying in particular that arrival $n_i + L$ observes an empty system, i.e., $W_{n_i+L} = 0$. The value of the remaining work remains within the set $[0,\ D_0]$ during this unloading period, leading to $W_{n_i+k} \le D_0$, for each $k = 1, \dots, L-1$. To explain the lower bound of the probability $\mathsf{P}(\mathcal{A}_i(L))$ in (5.1.21) below, note that, due to condition (5.1.4) and bound (5.1.18), we may write

$$\begin{aligned} \mathsf{P}\big(\tau_{n_i} > S_{n_i} + \delta,\ W_{n_i} \le D_0\big) &= \int_0^{D_0} \mathsf{P}(\tau_{n_i} > S_{n_i} + \delta |\ W_{n_i} = x)\ \mathsf{P}(W_{n_i} \in dx) \\ &= \int_0^{D_0} \mathsf{P}(\tau(x) > S(x) + \delta)\ \mathsf{P}(W_{n_i} \in dx) \\ &\ge \inf_{x \le D_0} \mathsf{P}(\tau(x) > S(x) + \delta) \ge \delta_0 \varepsilon , \end{aligned} \tag{5.1.20}$$

where $\varepsilon > 0$ satisfies (5.1.4) with $D = D_0$. As we mention above, the following key inclusion holds:

$$\{\tau_{n_i} > S_{n_i} + \delta,\ W_{n_i} \le D_0\} \subseteq \{W_{n_i+1} \le D_0\} ,$$

implying, due to (5.1.20), that

$$\begin{aligned} &\mathsf{P}\big(\tau_{n_i+1} > S_{n_i+1} + \delta,\ \tau_{n_i} > S_{n_i} + \delta,\ W_{n_i} \le D_0\big) \\ &= \mathsf{P}\big(\tau_{n_i+1} > S_{n_i+1} + \delta,\ \tau_{n_i} > S_{n_i} + \delta,\ W_{n_i+1} \le D_0,\ W_{n_i} \le D_0\big) \\ &= \int_0^{D_0} \mathsf{P}(\tau(x) > S(x) + \delta)\ \mathsf{P}(\tau_{n_i} > S_{n_i} + \delta,\ W_{n_i} \le D_0,\ W_{n_i+1} \in dx) \\ &\ge \varepsilon\, \mathsf{P}\big(\tau_{n_i} > S_{n_i} + \delta,\ W_{n_i} \le D_0,\ W_{n_i+1} \le D_0\big) \ge \delta_0 \varepsilon^2 . \end{aligned}$$

Continuing in this way, we finally obtain from (5.1.18) and (5.1.19) that

$$\mathsf{P}(\mathcal{A}_i(L)) \ge \delta_0 \Big[\inf_{x \le D_0} \mathsf{P}(\tau(x) > S(x) + \delta) \Big]^L \ge \delta_0\, \varepsilon^L . \qquad (5.1.21)$$

Consequently, a regeneration of the workload process occurs in the interval $[n_i,\ n_i + L]$ with a probability that is lower bounded by a positive constant $\delta_0\,\varepsilon^L$. Because the sequence $\{n_i\}$ is deterministic, and neither L nor $\delta_0\,\varepsilon^L$ depend on n_i and i, then $\mathsf{E}\theta < \infty$. □

Obviously, for practical purposes, one does not want to define interarrival and service time distribution functions for each real value x, and a natural setting describing a feedback mechanism between the workload process and the arrival and/or service process can be formulated by defining K thresholds

$$0 =: x_0 < x_1 < \cdots < x_K ,$$

such that, for all $x \in [x_k,\ x_{k+1})$, the random variables $S(x)$ are distributed as a single random variable S_k, and all $\tau(x)$ are distributed as a random variable τ_k, $k = 0, \ldots, K$. (We set $x_{K+1} = \infty$.) In this framework, the assumptions of Theorem 5.1 can be simplified and reformulated as

$$\max_{0 \le k \le K} (\,\mathsf{E}\tau_k,\ \mathsf{E}S_k) < \infty\,; \quad \min_{0 \le k \le K} \mathsf{P}(\tau_k > S_k) > 0\,; \quad \mathsf{E}S_K < \mathsf{E}\tau_K\,.$$

Note that in a more general setting in [3], different sets of the thresholds for the service and interarrival times are considered as well, however, this does not fundamentally alter the stability analysis.

Problem 5.1 Prove that (5.1.16) contradicts (5.1.17). (Also see Problem 2.1 in Sect. 2.1.)

In order to illustrate these results, consider the $GI/G/1$ system with interarrival time τ, service time S and with *impatient customers*. In particular, there exist iid *thresholds* $\{\gamma_n\}$ such that customer n leaves the system without service if $W_n > \gamma_n$ (which is called *balking*). The random variables $\{\gamma_n\}$ are assumed to be independent of the interarrival and service times. We also assume that a customer can wait in the queue 'infinitely long' with probability $\lim_{x\to\infty} \mathsf{P}(\gamma \ge x) = p$. Then the negative drift assumption (5.1.3) becomes

$$\lim_{x\to\infty} \mathsf{E}U(x) = \mathsf{E}S \lim_{x\to\infty} \mathsf{P}(\gamma \ge x) - \mathsf{E}\tau = p\ \mathsf{E}S - \mathsf{E}\tau < 0\,, \quad (5.1.22)$$

which can be written as $p\rho < 1$. As to a regeneration condition, we note that

$$\mathsf{P}(\tau > S\,\mathbf{1}(\gamma \ge x)) \ge \mathsf{P}(\tau > S)\,,$$

and it follows from assumption (5.1.4) in Theorem 5.1 that this system is positive recurrent if conditions $p\rho < 1$ and $\mathsf{P}(\tau > S) > 0$ hold. Condition (5.1.22) always holds if $p = 0$, and in particular if the acceptable waiting time in the system is upper bounded by a finite constant. If $p = 1$, then we obtain the classical system under condition $\rho < 1$, rendering condition $\mathsf{P}(\tau > S) > 0$ superfluous.

5.2 A Queue-Size Dependent System

In this section, maintaining the previous notations, we briefly consider a quite similar *queue-size dependent* $GI/G/1$ system again with impatient customers. This is a natural control mechanism, where a customer decides to either enter the system or leave upon arrival, provided the information about the queue size is available. We assume that an arriving customer k joins the system with probability p_n if he observes the number of customers in the system $\nu_k \le n$. Let us denote by $\limsup_{n\to\infty} p_n = p$, and consequently, p is an upper bound for the probability that a customer joins an 'infinitely large' queue. Evidently, this system regenerates when a new arrival meets it idle, and we denote by T the regeneration cycle length.

Theorem 5.2 *If* $\mathsf{P}(\tau > S) > 0$ *and* $p\rho < 1$, *then* $\mathsf{E}T < \infty$.

Proof The proof of this statement resembles the proof of Theorem 5.1, and we therefore merely outline it. First of all we refer to the statement of Theorem 2.6, showing in particular that $W_n \Rightarrow \infty$ implies $\nu_n \Rightarrow \infty$. (It is easy to verify that Theorem 2.6 holds for the system under consideration.) Define the indicator function $\mathsf{1}_n = 1$ if customer n joins the system, and $\mathsf{1}_n = 0$ otherwise. Then, using the Lindley recursion

$$W_{n+1} = (W_n + S_n\, \mathsf{1}_n - \tau_n)^+ \,, \quad n \ge 1 \,,$$

and denoting the increments by $\Delta_n = W_{n+1} - W_n$, we easily obtain the inequality

$$\mathsf{E}\Delta_n \le \mathsf{E}(S_n\, \mathsf{1}(W_n) - \tau_n)\, \mathsf{P}(W_n + S_n\, \mathsf{1}(W_n) \ge \tau_n) \,, \tag{5.2.1}$$

where we use the notation $1_n = \mathsf{1}(W_n)$, since $\mathsf{1}_n$ and W_n in general are dependent, while S_n, τ_n and W_n are independent random variables. Assume now that $W_n \Rightarrow \infty$, then

$$\lim_{n\to\infty} \mathsf{P}(W_n + S_n\, \mathsf{1}(W_n) \ge \tau_n) = 1 \,,$$

and, because $\nu_n \Rightarrow \infty$ as well, then it easily follows from (5.2.1) by assumption $p\rho < 1$ that

$$\limsup_{n\to\infty} \mathsf{E}\Delta_n = \mathsf{E}S \limsup_{n\to\infty} p_n - \mathsf{E}\tau = p\,\mathsf{E}S - \mathsf{E}\tau < 0 \,.$$

This negative drift condition implies that $W_n \not\Rightarrow \infty$, and, combined with the assumption $\mathsf{P}(\tau > S) > 0$, in turn implies positive recurrence. □

The previous analysis can be easily generalized to a queue-size dependent system with thresholds

$$0 =: x_0 < \cdots < x_M < \infty , \quad (5.2.2)$$

such that, if $\nu_n \in [x_i, x_{i+1})$, then the service time of customer n is selected from an iid sequence $\{S_n^{(i)}\}$, with generic element $S^{(i)}$ and $\mathsf{E}S^{(i)} < \infty$, $i = 0, \ldots, M$ ($x_{M+1} := \infty$). In such a setting, the service time of customer n is distributed as

$$\sum_{i=0}^{M} S_n^{(i)} \mathbf{1}(x_i \le \nu_n < x_{i+1}) , \; n \ge 1 .$$

The statement of Theorem 5.2 holds as well in this case, provided that condition $\mathsf{P}(\tau > S) > 0$ is replaced by

$$\min_{0 \le i \le M} \mathsf{P}(\tau > S^{(i)}) > 0 ,$$

and condition $p\rho < 1$ is replaced by condition $p\lambda \, \mathsf{E}S^{(M)} < 1$.

5.3 A System with Adaptable Service Speed

The system that we focus on in this section is adopted from [4]. Let us consider a zero initial state FIFO $GI/G/1$-type queueing system with iid interarrival times $\tau_n = t_{n+1} - t_n$, with rate $\lambda = 1/\, \mathsf{E}\tau$, and iid service times S_n, $n \ge 1$. Let V_n be the sojourn time of customer n. Assume that there are *thresholds* as in (5.2.2) such that, if $V_n \in [x_i, x_{i+1})$, then the *service rate* equals

$$r(V_n) = r_i \in (0, \infty) , \; i = 0, \ldots, M ,$$

i.e., the service rate depends on the accumulated work at a customer arrival instant. More precisely, when the amount of remaining work just *after an arrival* equals $x \in [x_i, x_{i+1})$, then the server works at a constant speed r_i until the next customer arrival. Such a system belongs to a class of queues with 'adaptable service speed' (see Chap. 6 in [4]), and the service speed is indeed defined by the sojourn time of a new customer arrival. (Such a scenario exists in some modern computer systems where, for instance, the runtime of a new job becomes known at its arrival instant [5].)

Because, for a fixed speed r, the remaining work decreases by the value $r\tau$ during interarrival time τ (as long as the work is positive), then the sojourn time sequence $\{V_n\}$ constitutes a Markov chain, obeying the following Lindley-type recursion [4, 6]:

$$V_{n+1} = (V_n - r(V_n)\tau_n)^+ + S_{n+1} , \; n \ge 1 .$$

It is easy to verify that a regeneration of the system takes place when a new arrival observes the system in an idle state. Denote by $\rho_M = \lambda\, \mathsf{E}S/r_M$.

Theorem 5.3 *Assume that* $\rho_M < 1$ *and that*

$$\min_{0\le i\le M} P(r_i\tau > S) > 0\,. \tag{5.3.1}$$

Then $\mathsf{E}T < \infty$.

Proof Denote by $\Delta_n = V_{n+1} - V_n$, $n \ge 1$, the incremental sojourn time process, then

$$\begin{aligned}\Delta_n &= V_n - r(V_n)\tau_n + S_{n+1} - V_n - (V_n - r(V_n)\tau_n)\,1(V_n < r(V_n)\tau_n)\\ &\le S_{n+1} - r(V_n)\tau_n + r(V_n)\tau_n\, 1(V_n < r(V_n)\tau_n) =: \sigma_n\,. \end{aligned}\tag{5.3.2}$$

Assume now that

$$V_n \Rightarrow \infty,\quad n \to \infty\,, \tag{5.3.3}$$

and note that then

$$\mathsf{P}(r(V_n) = r_M) = \mathsf{P}(V_n \ge x_M) \to 1\,. \tag{5.3.4}$$

Moreover, we can write

$$\mathsf{E}r(V_n) = \mathsf{E}(r(V_n)|V_n < x_M)\,\mathsf{P}(V_n < x_M) + r_M\,\mathsf{P}(V_n \ge x_M)\,,$$

and it is easy to deduce that, in view of (5.3.4),

$$\lim_{n\to\infty} \mathsf{E}r(V_n) = r_M\,. \tag{5.3.5}$$

It then follows from (5.3.2) and from $\rho_M < 1$, that

$$\begin{aligned}\limsup_{n\to\infty} \mathsf{E}\Delta_n &\le \lim_{n\to\infty} \mathsf{E}\sigma_n = \mathsf{E}S - r_M\,\mathsf{E}\tau + \max_i r_i \lim_{n\to\infty} \mathsf{P}\Big(\frac{V_n}{\max_i r_i} < \tau_n\Big)\\ &= \mathsf{E}S - r_M\,\mathsf{E}\tau = r_M\,\mathsf{E}\tau(\rho_M - 1) < 0\,,\end{aligned}\tag{5.3.6}$$

which contradicts assumption (5.3.3). Note that customer $n+1$ finds a lesser amount of remaining work in the system upon arrival than customer n if $r(V_n)\tau_n > S_n$, and as a result, these events allow to unload the system. Then, as many times before, using regeneration condition (5.3.1), we can readily prove that $\mathsf{E}\theta < \infty$. □

Problem 5.2 Prove (5.3.5).

5.4 Notes

The inclusion of dependencies between successive service and/or interarrival times in a queueing system in order to better reflect real-life effects makes such a system more flexible, and therefore potentially more realistic. One of the most relevant contemporary reasons to study state-dependent systems is the so-called *green computing* paradigm. The relentless high energy demand of modern data centers contrasts dramatically with their under-utilization during a considerable fraction of time. This strongly instigates hardware vendors to implement various technologies that allow the data center owners to reduce their power budget. These technologies encourage switching between the full capacity utilization (or even performance boost states), see e.g. [7], and the implementation of low-power states, either on the CPU level [8–10], or on the level of the machine and its components [11]. However, this operational solution may cause a performance degradation and, in particular, a temporary instability. Therefore, it is crucially important to find a power saving scenario for a non-full capacity utilization regime. This issue is closely related to the power optimization problem, which is discussed in a number of contributions as well, see for instance, [12–14]. Such a type of control is also used in high-performance clusters, where users are required to provide a runtime estimate for the submitted jobs [15].

The analysis presented in Sect. 5.1 can also be developed for an even much more general *multiserver* system, where the nth interarrival time not only depends on the waiting time W_n, but also on the *assigned server* for customer n, and on his service time S_n [3]. This system still maintains the Markov property of the workload process, although the negative drift and regeneration assumptions understandably become more complicated. The systems where service control is based on the workload process are not new in the available queueing literature, see for instance, the early contribution [16]. The system that we consider in Sect. 5.2 is well-known, and in particular, its stability analysis can be found in the influential book [17], which, among others, contains a profound discussion on state-dependent queueing systems, especially with respect to earlier related works. Another related paper [18] studies, under monotonicity assumptions, the stability region of an exponential multiclass system with queue-size depending service rates. Various state-dependent systems similar to those we study in Sect. 5.3 are also considered in the papers [19, 20]. The key element of the analysis in the mentioned works is the Markovian property of the workload process, which allows to describe the dependence by (a modified) Tákacs integro-differential equation, when the arrival process is Poisson (or state-dependent Poisson). A rather complete bibliography on state-dependent queues satisfying Lindley-type recursions (up to 2006) can be found in [21, 22] as well.

References

1. Meyn, S.P., Tweedie, R.L.: Markov Chains and Stochastic Stability. Springer, London (1993)
2. Billingsley, P.: Probability and Measure, 3rd edn. Wiley, New York (1995)
3. Morozov, E.: Stability analysis of a general state-dependent multiserver queue. J. Math. Sci. **200**(4), 462–472 (2014)
4. Bekker, R.: Queues with state-dependent rates. Eindhoven: Technische Universiteit Eindhoven, Proefschrift. ISBN 90-386-0734-2 (2005)
5. Feitelson, D.G.: Workload Modeling for Computer Systems Performance Evaluation. Cambridge University Press, Cambridge (2015)
6. Morozov, E., Rumyantsev, A.: A state-dependent control for green computing. In: Proceedings $30th$ International Symposium on Computer and Information Sciences (ISCIS 2015), Lecture Notes in Electrical Engineering, vol. 363, pp. 57–67 (2015)
7. Intel TurboBoost Technology http://www.intel.ru/content/www/ru/ru/architecture-and-technology/turbo-boost/turbo-boost-technology.html
8. Enhanced Intel SpeedStep Technology. http://www.intel.com/cd/channel/reseller/ASMO-NA/ENG/203838.htm]
9. IBM EnergyScale for POWER7 Processor-Based Systems. http://www-03.ibm.com/systems/power/hardware/whitepapers/energyscale7.html
10. Linux CPUFreq. https://www.kernel.org/doc/Documentation/cpu-freq/governors.txt
11. Advanced Configuration and Power Interface. https://www.acpi.info/spec.htm
12. Alonso, M., Martinez, J.-M., Santonja, V., Lopez, P., Duato, J.: Power saving in regular interconnection networks. Paral. Comp. **36**(12), 696–712 (2010)
13. Gandhi, A., Harchol-Balter, M., Das, R., Kephart, J.O., Lefurgy, C.: Power capping via forced idleness. In: Proceedings Workshop on Energy Efficient Design, pp. 1–6. Austin (2009)
14. Gandhi, A., Harchol-Balter, M., Das, R., Lefurgy, C.: Optimal power allocation in server farms. In: Proceedings ACM SIGMETRICS/Performance'09, pp. 157–168. Seattle (2009)
15. Tsafrir, D., Etsion, Y., Feitelson, D.G.: Modeling user runtime estimates. In: Proceedings Workshop on Job Scheduling Strategies for Parallel Processing (LNCS 3834), pp. 1–35. Vancouver (2005)
16. Callahan, J.R.: A queue with waiting time dependent service times. Nav. Res. Log. Quart. **20**(2), 321–324 (1973)
17. Gnedenko, B., Kovalenko, I.N.: An Introduction to Queueing Theory, 2nd edn. Birkhäuser, Basel (1989)
18. Borst, S., Jonckheere, M., Leskelä, L.: Stability of parallel queueing systems with coupled service rates. Discr. Ev. Dyn. Syst. **18**(4), 447–472 (2008)
19. Bekker, R.: Finite-buffer queues with workload-dependent service and arrival rates. Queueing Syst. **50**, 231–253 (2005)
20. Bekker, R., Borst, S.C., Boxma, O.J., Kella, O.: Queues with workload-dependent arrival and service rates. Queueing Syst. **46**, 537–556 (2004)
21. Boxma, O.J., Vlasiou, M., On queues with service and interarrival times depending on waiting times. Technical Report 2006-008, Eurandom, Eindhoven (2006)
22. Vlasiou, M.: Lindley-type recursions. Ph.D thesis, Eindhoven University of Technology, The Netherlands (2006)

Chapter 6
N-Models

In this chapter, we study the stability of a two-station queueing system with interacting servers that are able to join forces in one direction only, sometimes referred to as *N-models* [1–3]. Such a queueing system configuration has a lot in common with the notion of *flexible* servers [4], meaning that some service capacity may be transferred from one station, possibly with multiple servers, to another in order to accommodate varying demands.

First, in Sect. 6.1, we consider a *cascade* queueing network consisting of two coupled infinite-capacity single-server stations, where the 1st station has a Poisson arrival process (class-1 customers), and the input process of the 2nd station (class-2 customers) is renewal. Consecutive service times are modelled as iid, with a station-dependent distribution function. It is assumed that customers from station 1 may move to station 2 when the number of waiting class-1 customers exceeds a threshold. A transfer in the opposite direction however is not allowed. We would like to emphasize that in this model class-2 customers have priority over the migrated class-1 customers in station 2.

In Sect. 6.2, we study a multiserver variant of a two-station cascade N-model in which customers from station 1 may occupy servers of the 2nd station with *preemptive-resume* priority, when the queue size in the 1st station exceeds a threshold. Hence, in this model, migrated class-1 customers have priority in station 2 over class-2 customers.

As we show later on, the difference between the priority disciplines in these two models invokes a considerable difference in the respective stability conditions.

E. Morozov and B. Steyaert, *Stability Analysis of Regenerative Queueing Models*,
https://doi.org/10.1007/978-3-030-82438-9_6

6.1 A Two-Node Cascade System

6.1.1 Description of the Model

We consider a queueing system with two infinite-buffer single-server stations working in parallel, and class-i customers arriving to station i at the time instants $\{t_n^{(i)}\}$, which are assumed to form a renewal input process with iid interarrival times $\tau_n^{(i)} = t_{n+1}^{(i)} - t_n^{(i)}$, $n \ge 1$, with rate $\lambda_i = 1/\mathsf{E}\tau^{(i)}$, $i = 1, 2$. We further assume that the generic interarrival time $\tau^{(1)}$ is *exponential*, while $\tau^{(2)}$ has a general distribution. It is also assumed that the service times $\{S_n^{(i)}\}$ at station i are iid with rate $\mu_i = 1/\mathsf{E}S^{(i)}$, $i = 1, 2$. If there are more than $\mathbb{C}$ class-1 customers in the queue of station 1 and station 2 becomes empty, a customer from the 1st station switches to the 2nd one and starts service immediately, becoming a *class-(1, 2) customer*. For the subsequent stability analysis it does not matter which customer from station 1 makes this jump. Class-(1, 2) customers are assumed to have iid service times $\{S_n^{(12)}\}$ with rate $\mu_{12} = 1/\mathsf{E}S^{(12)}$. In station 2, class-2 customers have *non-preemptive priority* over class-(1, 2) customers, in the sense that jumps can only occur if station 2 is free of class-2 customers, as depicted in Fig. 6.1, where class-i customers are denoted by C_i.

Let $W_i(t)$ and $\nu_i(t)$ be the remaining work and the number of customers, respectively, in station i at instant t. Define the continuous-time workload process

$$\mathbf{W}(t) = (W_1(t),\ W_2(t)),\ t \ge 0,$$

and the discrete-time workload process embedded at the arrival instants of *class-2 customers*

$$W_k^{(i)} = W_i(t_k^{(2)-}),\ i = 1, 2;\quad \mathbf{W}_k = (W_k^{(1)},\ W_k^{(2)})\ ,\ k \ge 1\ .$$

The regeneration instants of the process $\{\mathbf{W}(t)\}$ are now defined as follows:

$$T_{n+1} = \min_{k \ge 1}(t_k^{(2)} > T_n : \mathbf{W}_k = \mathbf{0}),\ n \ge 0;\ T_0 := 0\ , \tag{6.1.1}$$

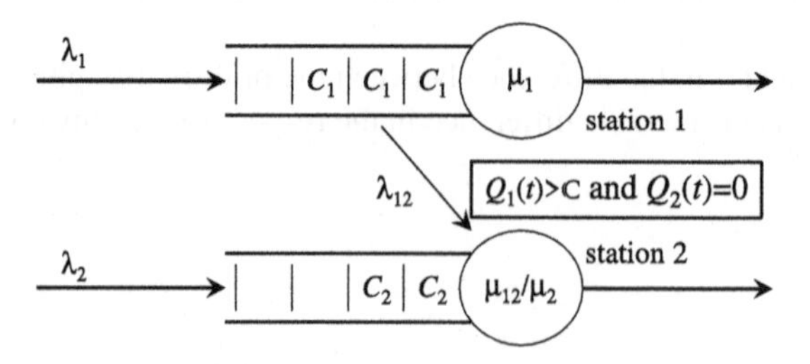

Fig. 6.1 A cascade network with two stations; priority for class-2 customers

while the regeneration instants of the process $\{\mathbf{W}_k\}$ are given by

$$\theta_{n+1} = \min_{k \geq 1}(k > \theta_n : \mathbf{W}_k = \mathbf{0}), \ n \geq 0; \ \theta_0 = 1 , \tag{6.1.2}$$

meaning that a regeneration of the system occurs when a class-2 customer observes the entire system as being empty. Note that in (6.1.1) and (6.1.2), we have exploited the memoryless property of the interarrival time $\tau^{(1)}$ of the Poisson input to station 1 by replacing, at the instant $t_k^{(2)}$, the remaining time $\tau^{(1)}(t_k^{(2)})$ up to the next class-1 arrival by an independent copy of $\tau^{(1)}$, see also Remark 2.1.

Remark 6.1 The analysis in this chapter can be readily extended to a similar system with a common renewal input (with renewal instants $\{t_n\}$), in which an arbitrary customer is a class-i one with probability p_i, regardless of the state of the system, $p_1 + p_2 = 1$. Such a setting is considered in some of the previous chapters, and in Sect. 6.2 as well. In such a case, the instant $t_k^{(2)}$ in the above definitions must be replaced by t_k.

6.1.2 Stability Analysis

First of all, we establish a preliminary result, upon which we rely in the following analysis. Consider a modification of the basic model, in which the workload process maintains the same distribution function. We focus on station 2, and will adopt a *preemptive resume* priority discipline, whereby the service of a class-(1, 2) customer is interrupted by newly arriving class-2 customers. It is easy to see that the amount of type-2 work that is processed in between jumps *remains unchanged* under such a transformation because jumps from station 1 to station 2 only occur at instants t where $W_2(t) = 0$. The workload processes in both systems are asymptotically equivalent (this is sufficient for the analysis that follows), and we will focus on this modification of the original system, unless mentioned otherwise. In the modified system, denote by $\{\widehat{W}_2(t)\}$ the remaining work at instant t generated by class-2 customers, and define $\widehat{W}_k^{(2)} = \widehat{W}_2(t_k^{(2)-})$. The regeneration instants of the process $\{\widehat{W}_2(t)\}$ are then given by

$$T_{n+1}^{(2)} = \min(t_k^{(2)} > T_n^{(2)} : \widehat{W}_k^{(2)} = 0) \ , \ n \geq 0 \ , \ T_0^{(2)} := 0 \ , \tag{6.1.3}$$

i.e., , these instants occur when newly arriving class-2 customers see no other class-2 customers in station 2.

Theorem 6.1 *If condition*

$$\rho_2 := \lambda_2/\mu_2 < 1 \ , \tag{6.1.4}$$

is satisfied, then the workload process $\{W_2(t)\}$ is tight under any initial state.

Proof The proof is based on the following observation. In view of condition (6.1.4), the process $\{\widehat{W}_2(t)\}$ is positive recurrent regenerative, and therefore *tight* for any initial state, which follows from Theorem 2.1. Now we show that the process $\{S_{12}(t)\}$, describing the remaining service time of class-(1, 2) customers in the original system is a tight process. Consider a modification of the original station 2, where class-(1, 2) customers arrive as long as station 2 is empty of class-2 customers. Because of the preemptive-resume priority service discipline, one can perceive the system as if we 'discard' the time periods during which station 2 serves class-2 customers, and hence this new time scale only takes into account the time that can be devoted to serving class-(1, 2) customers. Thus, for each value of t, the input of class-(1, 2) customers to station 2 is a renewal process over a time period with length

$$\widehat{I}_2(t) := \int_0^t 1(\widehat{W}_2(u) = 0)du , \tag{6.1.5}$$

when station 2 is free of class-2 customers. This construction is then applied, as in Theorem 3.3, to establish the tightness of the remaining service time process $\{S_{12}(t)\}$, where we also take into account that if a class-(1, 2) customer is being served, then interruptions initiated by new class-2 arrivals are generated independently of this service process. Finally,

$$W_2(t) = \widehat{W}_2(t) + S_{12}(t) ,$$

and the conclusion of the theorem readily follows. □

Remark 6.2 Unlike the process $\{\widehat{W}_2(t)\}$, we can not claim that the process $\{W_2(t)\}$ is positive recurrent regenerative under assumption (6.1.4), due to the *interaction between both stations*.

Now we will prove the following main result.

Theorem 6.2 *Assume that the following condition*

$$\rho_2 + \frac{(\lambda_1 - \mu_1)^+}{\mu_{12}} < 1 , \tag{6.1.6}$$

is satisfied. Then $\mathsf{E}T < \infty$.

First, it is convenient to split condition (6.1.6) into two separate ones, yielding

$$\rho_2 < 1 \quad \text{and} \quad \lambda_1 < \mu_1 ; \tag{6.1.7}$$

$$\lambda_1 + \rho_2\mu_{12} < \mu_1 + \mu_{12} \quad \text{and} \quad \lambda_1 \geq \mu_1 . \tag{6.1.8}$$

Condition (6.1.7) corresponds to a situation when each of the two stations, working individually, has sufficient capacity to serve the arriving work. Condition (6.1.8) indicates that if the 1st station is overloaded, then the system works as a two-server system with input rate $\lambda_1 + \lambda_2\mu_{12}/\mu_2$, and with total service rate $\mu_{12} + \mu_1$. Condition (6.1.8) becomes intuitively clear when $\mu_{12} = \mu_2$, resulting in the condition

$$\lambda_1 + \lambda_2 < \mu_1 + \mu_2 .$$

We also split the proof into two parts according to these two cases.

Theorem 6.3 *If assumption (6.1.7) holds, then* $\mathsf{E}T < \infty$.

Proof The potential trickling down of class-1 customers to station 2 can only reduce the work in the original station 1 in comparison with station 1 in *isolation*, with a workload process denoted by $\{\widehat{W}_1(t)\}$, which is *positive recurrent* regenerative by Theorem 2.1. The remainder of the proof is then based on the coupling of service times and arrival times, and the well-known monotonicity property of the workload process in the single-server $M/G/1$ system. Indeed, we set the service time $S_n^{(1)} = 0$ if class-1 customer n becomes a class-(1, 2) customer, then $W_1(t) \le \widehat{W}_1(t)$ for all t by Theorem 3.8. Note that the workload process $\{\widehat{W}_1(t)\}$ is tight, and moreover,

$$\liminf_{t\to\infty} \mathsf{P}(W_1(t) = 0) \ge \lim_{t\to\infty} \mathsf{P}(\widehat{W}_1(t) = 0) = 1 - \lambda \mathsf{E}S^{(1)} = 1 - \rho_1 =: \varepsilon > 0 . \quad (6.1.9)$$

Invoking Theorem 6.1 the process $\{W_2(t)\}$ is tight, and consequently, using (6.1.9), we can choose a constant D such that

$$\mathsf{P}\big(W_1(t) = 0,\ W_2(t) \le D\big) \ge \varepsilon/2 , \quad (6.1.10)$$

for all sufficiently large values of t. Note that the assumption $\rho_2 < 1$ also implies the regeneration condition

$$\mathsf{P}(\tau^{(2)} > S^{(2)}) > 0 . \quad (6.1.11)$$

Finally, using the latter condition, the exponentiality of $\tau^{(1)}$ and (6.1.10), we may unload the system in a standard way and obtain a regeneration in a finite time with a positive probability, i.e., $\mathsf{E}T < \infty$. □

Next, let us focus on the more challenging case (6.1.8).

Theorem 6.4 *If assumption (6.1.8) is satisfied, then* $\mathsf{E}T < \infty$.

Proof First of all, note that we may write assumption (6.1.8) as

$$\rho_2 < 1 + \frac{\mu_1 - \lambda_1}{\mu_{12}} \le 1 ,$$

and thus regeneration condition (6.1.11) holds in this case as well. Note that, at instant t, a jump from station 1 to station 2 is possible only if the number of customers at station 1 satisfies $\nu_1(t) \ge \mathbb{C} + 1$. Denote by

$$I_1(t) = \int_0^t 1(\nu_1(u) = 0)du ,$$

the idle time of server 1 in the time interval $[0,\ t]$. Also define

$$J_1(t) = \int_0^t 1(\nu_1(u) > \mathbb{C})du ,$$

the amount of time during which there are class-1 customers which can jump from station 1 to station 2, in the interval $[0,\ t]$. Also note that $\widehat{I}_2(t)$ is the time (within $[0,\ t]$) when the 2nd station is either idle or only contains a class-(1, 2) customer.

Before writing down the balance equation below, we note that, because of the two distinct input processes, we do not assume that under an arbitrary initial state of the whole system the time $t = 0$ is an arrival instant. Now we obtain the following balance equation describing the dynamics of the 1st station:

$$W_1(0^-) + V_1(t) = W_1(t) + L_1(t) + t - I_1(t) , \tag{6.1.12}$$

where $V_1(t)$ is the amount of work that has arrived in the 1st station, and $L_1(t)$ the amount of work lost in station 1 because of the migration of customers from station 1, in the time interval $[0,\ t]$. Denote by $A_1(t)$ the number of class-1 arrivals in $[0,\ t]$, and by 1_k the indicator function such that $1_k = 1$ if the kth class-1 customer becomes a class-(1, 2) one. Also denote by $\mathcal{L}(t)$ the set of indices of the class-1 customers who become class-(1, 2) customers in the time interval $[0,\ t]$. Then we can write

$$V_1(t) = \sum_{k=1}^{A_1(t)} S_k^{(1)} , \tag{6.1.13}$$

$$L_1(t) = \sum_{k=1}^{A_1(t)} 1_k S_k^{(1)} = \sum_{k \in \mathcal{L}(t)} S_k^{(1)} =_{st} \sum_{k=1}^{A_{12}(t)} S_k^{(1)} , \tag{6.1.14}$$

where $A_{12}(t)$ represents the number of class-(1, 2) customer arrivals in station 2 during $[0,\ t]$. Define the sets

$$\mathcal{J}_1(t) = \{u : \nu_1(u) > \mathbb{C},\ u \in [0,\ t]\} ,\ \mathcal{I}_2(t) = \{u : \widehat{W}_2(u) = 0,\ u \in [0,\ t]\} .$$

Then the Lebesgue measures $|\cdot|$ of these sets are respectively given by

$$|\mathcal{J}_1(t)| = J_1(t) ,\quad |\mathcal{I}_2(t)| = \widehat{I}_2(t) ,\quad t \geq 0 .$$

Note that

$$|\mathcal{J}_1(t) \cap \mathcal{I}_2(t)| = \int_0^t 1(\nu_1(u) > \mathbb{C},\ \widehat{W}_2(u) = 0)du ,$$

is the time, within the interval $[0,\ t]$, during which class-1 customers jump from station 1 to station 2 becoming class-(1, 2) customers. Denote by

$$\int_0^t 1(\nu_1(u) \le \mathbb{C})du = J_{\mathbb{C}}(t),\ t \ge 0\,, \tag{6.1.15}$$

and note that $J_{\mathbb{C}}(t) + J_1(t) = t$. In view of

$$\widehat{I}_2(t) \le \int_0^t 1(\nu_1(u) > \mathbb{C},\ \widehat{W}_2(u) = 0)du + J_{\mathbb{C}}(t)\,,$$

we obtain the following inequalities:

$$\widehat{I}_2(t) - J_{\mathbb{C}}(t) \le |\mathcal{J}_1(t) \cap \mathcal{I}_2(t)| \le \widehat{I}_2(t)\,. \tag{6.1.16}$$

Assume now that

$$\nu_1(t) \Rightarrow \infty\,,\ t \to \infty\,. \tag{6.1.17}$$

Then we find

$$\lim_{t\to\infty} \frac{1}{t} \int_0^t \mathsf{P}(\nu_1(u) \le \mathbb{C})du = \lim_{t\to\infty} \frac{1}{t} \mathsf{E} J_{\mathbb{C}}(t) = 0\,. \tag{6.1.18}$$

Because of the assumption $\rho_2 < 1$ and the preemptive-resume priority service discipline, only the 2nd station is positive recurrent regenerative. (At this point, we again rely on definition (6.1.3) of the regeneration instances of the 2nd station.) This implies in particular

$$\lim_{t\to\infty} \frac{\widehat{I}_2(t)}{t} = 1 - \rho_2\,, \quad \text{w.p.1}\,,$$

and from the dominated convergence theorem we can also deduce that

$$\lim_{t\to\infty} \frac{\mathsf{E}\widehat{I}_2(t)}{t} = 1 - \rho_2 > 0\,. \tag{6.1.19}$$

It then follows from (6.1.16), (6.1.18) and (6.1.19) that

$$\lim_{t\to\infty} \frac{1}{t} \mathsf{E}|\mathcal{J}_1(t) \cap \mathcal{I}_2(t)| = \lim_{t\to\infty} \frac{1}{t} \int_0^t \mathsf{P}(\nu_1(u) > \mathbb{C},\ \widehat{W}_2(u) = 0)du = 1 - \rho_2\,.$$

Since summation index $A_{12}(t)$ and the summands $S_k^{(1)}$ in (6.1.14) are independent, this leads to

$$\mathsf{E}L_1(t) = \mathsf{E}\sum_{k=1}^{A_{12}(t)} S_k^{(1)} = \sum_{k=1}^{\infty} \mathsf{E}\big[S_k^{(1)} 1(A_{12}(t) \ge k)\big]$$
$$= \mathsf{E}S^{(1)} \sum_{k=1}^{\infty} \mathsf{P}(A_{12}(t) \ge k) = \frac{\mathsf{E}A_{12}(t)}{\mu_1} . \tag{6.1.20}$$

By the same reason, again using Wald's identity, $\mathsf{E}V_1(t) = \mathsf{E}A_1(t)/\mu_1$, and invoking the elementary renewal theorem (1.2.11), we obtain

$$\lim_{t\to\infty} \frac{1}{t}\mathsf{E}V_1(t) = \frac{\lambda_1}{\mu_1} = \rho_1 . \tag{6.1.21}$$

Note that, in view of assumption (6.1.17),

$$\lim_{t\to\infty} \frac{\mathsf{E}I_1(t)}{t} = \lim_{t\to\infty} \frac{1}{t}\int_0^t \mathsf{P}(\nu_1(u) = 0)du = 0 . \tag{6.1.22}$$

Now taking into account (6.1.20)–(6.1.22) we obtain from (6.1.12) the following relation:

$$\rho_1 = \limsup_{t\to\infty} \frac{\mathsf{E}W_1(t)}{t} + \frac{1}{\mu_1}\limsup_{t\to\infty} \frac{\mathsf{E}A_{12}(t)}{t} + 1 . \tag{6.1.23}$$

Next, consider in more detail the 2nd term in the right-hand side of expression (6.1.23). Let $Z_{12}(t)$ be the number of renewals in the *renewal process* generated by the service times of class-(1, 2) customers realized in station 2 in the interval $[0, t]$. Then, due to the elementary renewal theorem,

$$\lim_{t\to\infty} \frac{1}{t}\mathsf{E}Z_{12}(t) = \mu_{12} . \tag{6.1.24}$$

Due to the specifics of the service discipline, $Z_{12}(\widehat{I_2}(t))$ is an upper bound of the actual number $A_{12}(t)$ of class-(1, 2) customers which arrive to station 2 in $[0, t]$. Consequently, we may write

$$A_{12}(t) \le Z_{12}(\widehat{I_2}(t)),\ t \ge 0 . \tag{6.1.25}$$

Moreover, because the service times $\{S_n^{(12)}\}$ and time $\widehat{I_2}(t)$ are independent, then conditioned on the event $\{\widehat{I_2}(t) = x\}$, the quantity $Z_{12}(\widehat{I_2}(t))$ is distributed as $Z_{12}(x)$. From (6.1.19) and (6.1.24), we then find

$$\lim_{t\to\infty} \frac{1}{t}\mathsf{E}Z_{12}(\widehat{I_2}(t)) = \lim_{t\to\infty} \frac{1}{t}\int_1^t \frac{\mathsf{E}Z_{12}(u)}{u}\, u\, \mathsf{P}(\widehat{I_2}(t) \in du)$$
$$= \mu_{12} \lim_{t\to\infty} \frac{1}{t}\mathsf{E}\widehat{I_2}(t) = \mu_{12}(1-\rho_2) . \tag{6.1.26}$$

On the other hand, it is clear that the difference $Z_{12}(\widehat{I_2}(t)) - A_{12}(t)$ is upper bounded by the number of 'potential' jumps from station 1 which could occur during time periods when $\nu_1(u) \le \mathbb{C}$, $u \in [0, t]$, and the total length of this time is $J_{\mathbb{C}}(t)$. Because of (6.1.18), $\mathsf{E}J_{\mathbb{C}}(t) = o(t)$, and it follows from renewal theory that the time-average fraction of such potential jumps must be $o(t)$ as well, implying by (6.1.25)

$$\lim_{t\to\infty} \frac{1}{t}\mathsf{E}\big[Z_{12}(\widehat{I_2}(t)) - A_{12}(t)\big] = 0 .$$

It then follows from (6.1.26) that

$$\lim_{t\to\infty} \frac{1}{t}\mathsf{E}A_{12}(t) = \mu_{12}(1 - \rho_2) . \tag{6.1.27}$$

Expressions (6.1.23) and (6.1.27) then result in

$$\limsup_{t\to\infty} \frac{\mathsf{E}W_1(t)}{t} = \rho_1 - \frac{\mu_{12}}{\mu_1}(1 - \rho_2) - 1 \ge 0 .$$

(Note that, alternatively, we could operate with lim inf as well.) We then finally find the following inequality:

$$\lambda_1 + \mu_{12}\rho_2 \ge \mu_1 + \mu_{12} ,$$

which is opposite to condition (6.1.8). This contradiction shows that assumption (6.1.17) is false, and there exists a deterministic sequence of time instants $z_i \to \infty$ and constants C, $\varepsilon > 0$ such that

$$\inf_i \mathsf{P}(\nu_1(z_i) \le C) \ge \varepsilon . \tag{6.1.28}$$

Taking into account the tightness of the process $\{W_2(t)\}$ and the exponentiality of the interarrival time $\tau^{(1)}$, we can now readily establish that, starting at some instant z_i, and in view of the bound (6.1.28), the process $\{\nu_1(t),\ W_2(t)\}$, and hence, the process $\{\mathbf{W}(t)\}$ as well hits the regeneration state with a positive probability in a finite interval, where both the probability and the length of the interval do not depend on z_i and i. Hence, the proof is completed. □

The previous analysis can be readily extended to a generalized system with $N_1 > 1$ servers at station 1. Namely, the following statement holds [5]:

Theorem 6.5 *If, in the generalized system, condition*

$$\rho_2 + \frac{(\lambda_1 - N_1\mu_1)^+}{\mu_{12}} < 1 ,$$

is satisfied, then $\mathsf{E}T < \infty$.

6.1.3 Instability

In this section, we focus on the original system with two single-server stations, and show that the sufficient stability condition (6.1.6) is indeed the stability criterion, for the zero initial state case where the 1st class-2 customer arrives in an empty system. For the purpose of an instability analysis, then in view of the results presented in the previous sections, it is natural to consider the following condition:

$$\mu_{12}\rho_2 + (\lambda_1 - \mu_1)^+ \geq \mu_{12} \,. \tag{6.1.29}$$

Moreover, throughout this section we assume that

$$\mathsf{P}(\tau^{(2)} > S^{(2)}) > 0 \,. \tag{6.1.30}$$

Theorem 6.6 *If conditions (6.1.29) and (6.1.30) hold, then* $\mathsf{E}T = \infty$.

Proof Assume that $\lambda_1 \leq \mu_1$. Then (6.1.29) reduces to $\rho_2 \geq 1$, and it is well-known that $\widehat{W}_2(t) \Rightarrow \infty$ even in the isolated station 2, that is $\mathsf{E}T = \infty$. Assume now that $\lambda_1 > \mu_1$, in which case condition (6.1.29) becomes

$$\mu_{12}\rho_2 + \lambda_1 \geq \mu_1 + \mu_{12} \,. \tag{6.1.31}$$

Denote by $I_2(t)$ the idle time of station 2 and by $V_{12}(t)$ the amount of work which class-(1, 2) customers bring in station 2, during the interval $[0, t]$. We obtain the following equation describing the dynamics of the 2nd station

$$V_2(t) + V_{12}(t) = W_2(t) + t - I_2(t) \,, \quad t \geq 0 \,. \tag{6.1.32}$$

Assume that $\mathsf{E}T < \infty$, then $W_2(t) = o(t)$ and all limits w.p.1 below exist. In particular, $\{A_{12}(t)\}$ is a *non-decreasing* process with regenerative increments in which the mean (cycle) increment is upper bounded by $\lambda_1 \mathsf{E}T$, the mean number of class-1 arrivals within a regeneration cycle. So (1.2.7) implies that the limit

$$\lambda_{12} = \lim_{t\to\infty} \frac{1}{t} A_{12}(t) \,, \tag{6.1.33}$$

exists, which is the arrival rate of class-(1, 2) customers. Because of

$$\lim_{t\to\infty} \frac{1}{t} V_{12}(t) = \lim_{t\to\infty} \frac{1}{t} \sum_{n=1}^{A_{12}(t)} S_n^{(12)} = \lambda_{12} \mathsf{E}S^{(12)} \,,$$

then dividing both sides of expressions (6.1.12) and (6.1.32) by t and letting $t \to \infty$, we obtain the two following equations

$$\rho_1 = \frac{\lambda_{12}}{\mu_1} + 1 - \mathsf{P}_0^{(1)}\,,$$
$$\rho_2 + \frac{\lambda_{12}}{\mu_{12}} = 1 - \mathsf{P}_0^{(2)}\,, \tag{6.1.34}$$

where

$$\mathsf{P}_0^{(i)} = \lim_{t\to\infty} \frac{I_i(t)}{t} = \frac{\mathsf{E}I_0^{(i)}}{\mathsf{E}T}\,,$$

is the stationary idle probability of station i and $I_0^{(i)}$ is the idle time of station i during a regeneration period, $i = 1, 2$. To show that $\mathsf{P}_0^{(i)} > 0$, we denote by I_0 the time period during which *both stations are idle* within a regeneration cycle, so $I_0 \leq_{st} I_0^{(i)}$. Note that a new regeneration cycle, which always starts by a class-2 customer arrival, may contain only this one class-2 customer with probability

$$\mathsf{P}\big(\min(\tau^{(1)}, \tau^{(2)}) > S^{(2)}\big) = \mathsf{P}(\tau^{(2)} > S^{(2)}) \int_0^\infty e^{-\lambda_1 x} F_2(dx) > 0\,,$$

where we use condition (6.1.30), and F_2 represents the distribution function of $S^{(2)}$. Moreover, again by relying on (6.1.30) and $\mathsf{E}S^{(2)} < \infty$, there exist constants $\delta > 0,\ \delta_0 > 0$ such that

$$\mathsf{P}(\tau^{(2)} > S^{(2)} + \delta) \geq \delta_0\,,$$

implying

$$\begin{aligned}\mathsf{E}I_0 &\geq \mathsf{E}(I_0;\ I_0 \geq \delta) = \mathsf{E}(I_0 | I_0 \geq \delta)\mathsf{P}(I_0 \geq \delta)\\ &\geq \delta \mathsf{P}\big(\tau^{(1)} > S^{(2)} + \delta,\ \tau^{(2)} > S^{(2)} + \delta\big)\\ &\geq \delta\delta_0 \int_0^\infty e^{-\lambda_1(x+\delta)} F_2(dx) > 0\,.\end{aligned} \tag{6.1.35}$$

It then follows that

$$\mathsf{P}_0^{(i)} \geq \frac{\mathsf{E}I_0}{\mathsf{E}T} > 0\,,\ i = 1, 2\,.$$

By summation of the equations in (6.1.34), we obtain the inequality

$$\rho_2\mu_{12} + \lambda_1 = \mu_1 + \mu_{12} - \mu_1\mathsf{P}_0^{(1)} - \mu_{12}\mathsf{P}_0^{(2)} < \mu_1 + \mu_{12}\,,$$

which contradicts (6.1.31), and thus $\mathsf{E}T = \infty$. □

Consequently, condition (6.1.6) is the *stability criterion* of the cascade system with zero initial state.

Problem 6.1 Explain the chain of inequalities in (6.1.35).

6.2 The Two-Station N-Model

In this section we consider a similar system with interacting stations, the difference being that class-(1, 2) customers now have *preemptive-resume priority* in station 2, as depicted in Fig. 6.2. In this alternative system, *throughput maximisation* is in general not attainable because a part of the service capacity at station 1 remains unused, and this implies a radical difference with the stability conditions of the previous cascade system.

6.2.1 Model Description

Mainly following [6], we consider a two-station queueing system with N_1 and N_2 parallel servers respectively, which is fed by a general renewal input process with rate λ and two classes of customers: with probability p_i a new customer belongs to class i and is directed to station i, regardless of the state of system, $p_1 + p_2 = 1$. We define $Q_i(t)$ as the the *queue size* (the number of customers waiting for service, not including those being served) in station i at instant t, $i = 1, 2$. If $Q_1(t) \geq \mathbb{C}$, where $\mathbb{C}$ is a given threshold, and the number of class-(1, 2) customers in station 2 (that have jumped from station 1, as explained above) is less than N_2 (implying that there exists at least one server which is free of class-(1, 2) customers), then a class-1 customer jumps to station 2, becoming a class-(1, 2) customer, and his service immediately starts. In other words, it is as if we apply *preemptive class-(1, 2) priority* in station 2, provided that $Q_1(t) \geq \mathbb{C}$. On the other hand, customer jumps from station 2 to station 1 are not allowed. As in the cascade system, any waiting class-1 customer may jump to station 2, if $Q_1(t) \geq \mathbb{C}$.

This model is a variation of the cascade system considered in Sect. 6.1 and is also called the N-*model* [1, 2]. In view of the above construction, the external input process in station i is a renewal process with rate $\lambda_i = \lambda\, p_i$, $i = 1, 2$. We will make a distinction between regenerations of the system as a whole, and regenerations of the 1st station, which are generated by class-1 customers arriving in an empty station 1, with generic regeneration period length $T^{(1)}$. More precisely, denoting by $\{t_n\}$ the

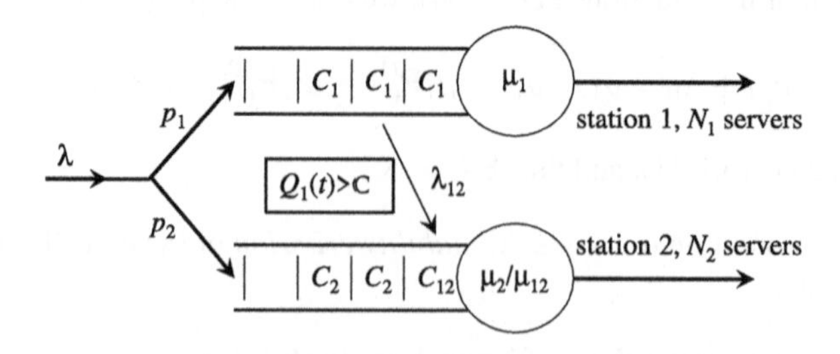

Fig. 6.2 N-model with two stations; priority for class-(1, 2) customers

arrival instants, $\nu_1(t_n^-) = \nu_n^{(1)}$, $\nu_{12}(t_n^-) = \nu_n^{(12)}$, and by $1_n^{(1)}$ the indicator function such that $1_n^{(1)} = 1$ if the nth arrival is class-1 customer (and $1_n^{(1)} = 0$ otherwise), then we define the regeneration instants of the 1st station, in continuous and discrete time respectively, as follows:

$$\begin{aligned} T_{n+1}^{(1)} &= \inf(t_k > T_n^{(1)} : \nu_k^{(1)} = 0,\ \nu_k^{(12)} = 0,\ 1_k^{(1)} = 1),\ n \geq 0\,;\ T_0^{(1)} := 0\,, \\ \theta_{n+1}^{(1)} &= \inf(k > \theta_n^{(1)} : \nu_k^{(1)} = 0,\ \nu_k^{(12)} = 0,\ 1_k^{(1)} = 1),\ n \geq 0\,;\ \theta_0^{(1)} := 1\,, \end{aligned} \tag{6.2.1}$$

with generic regeneration period lengths $T^{(1)}$ and $\theta^{(1)}$ respectively. Note that the requirement $\nu_k^{(12)} = 0$ reflects a dependence of the states of station 1 on the presence/absence of class-(1, 2) customers in station 2.

Remark 6.3 Assume for a moment that this N-system has the same input process as the cascade system considered in the previous section, i.e., $\tau^{(1)}$ is exponential and $\tau^{(2)}$ has a general distribution. Then the event $\{1_k^{(1)} = 1\}$ in definition (6.2.1) could be omitted, since an arrival of a class-2 customer, when station 1 is idle and station 2 contains no class-(1, 2) customers, is still a regeneration instant of the 1st station. This is because the remaining interarrival time *in station 1* is exponential. (In this regard, also see Remark 2.1.) This case illustrates that in general there may exist various types of classical regenerations in a system, which are not always related to the events occurring in this system.

6.2.2 *Necessary Stability Conditions*

Stability of the 1st station. Assume positive recurrence, $\mathsf{E}T^{(1)} < \infty$, $T_1^{(1)} < \infty$, and denote the departed amount of work from station 1 during the time interval $[0, t]$ by

$$B_1(t) = \sum_{i=1}^{N_1} B_1^{(i)}(t) = tN_1 - \sum_{i=1}^{N_1} I_1^{(i)}(t) = tN_1 - I_1(t)\,,$$

where $B_1^{(i)}(t)$ and $I_1^{(i)}(t)$ represent the busy and idle time of server i respectively. Also define the sum of the remaining works in all servers of station 1,

$$\mathsf{W}_1(t) = \sum_{i=1}^{N_1} W_1^{(i)}(t)\,,\quad t \geq 0\,.$$

Now we obtain the following relation:

$$\begin{aligned} \mathsf{W}_1(0^-) + V_1(t) &= \mathsf{W}_1(t) + L_1(t) + B_1(t) \\ &= \mathsf{W}_1(t) + L_1(t) + N_1 t - I_1(t)\,,\ t \geq 0\,, \end{aligned} \tag{6.2.2}$$

where $L_1(t)$ is the *lost work* in station 1 defined in (6.1.14). Note that $\{L_1(t)\}$ is a (non-decreasing) positive recurrent process with regenerative increments,

$$\Delta_n := L_1(T_{n+1}^{(1)}) - L_1(T_n^{(1)}) \ , \ \ n \geq 1 \ ,$$

and a generic increment is upper bounded by $T^{(1)} N_1$. Then (1.2.14) and the SLLN imply that the following limit exists w.p.1:

$$\lim_{t\to\infty} \frac{1}{t} L_1(t) = \lim_{t\to\infty} \frac{\sum_{k\in\mathcal{L}(t)} S_k^{(1)}}{A_{12}(t)} \frac{A_{12}(t)}{A_1(t)} \frac{A_1(t)}{t} = \rho_1 \mathsf{P}_\ell \ , \tag{6.2.3}$$

where, by (6.1.33), the limit

$$\mathsf{P}_\ell := \lim_{t\to\infty} \frac{A_{12}(t)}{A_1(t)} = \lim_{t\to\infty} \frac{A_{12}(t)/t}{A_1(t)/t} = \frac{\lambda_{12}}{\lambda_1} \ , \tag{6.2.4}$$

is equal to the stationary probability that a class-1 customer jumps from station 1 to station 2, implying the expected equality $\lambda_{12} = \lambda_1 \mathsf{P}_\ell$. Let $\mathsf{P}_B^{(i)}$ be the stationary busy probability of server i (in station 1), i.e.,

$$\lim_{t\to\infty} \frac{B_1^{(i)}(t)}{t} = \lim_{t\to\infty} \frac{1}{t} \int_0^t \mathsf{1}(W_1^{(i)}(u) > 0) du = \mathsf{P}_B^{(i)} \ .$$

Because $B_1(t)/t \leq N_1$, then due to the positive recurrence the limit

$$\lim_{t\to\infty} \frac{1}{t} B_1(t) = \sum_{i=1}^{N_1} \mathsf{P}_B^{(i)} =: \mathsf{E}\mathcal{B}_1 \ , \tag{6.2.5}$$

exists, and is equal to the *mean stationary number of busy servers* in the 1st station. Note that if the servers are identical, then $\mathsf{P}_B^{(i)} \equiv \mathsf{P}_B$, $\mathsf{E}\mathcal{B}_1 = N_1 \, \mathsf{P}_B$. Now, dividing both sides of (6.2.2) by t, letting $t \to \infty$, and using (6.2.3)–(6.2.5), we obtain $\rho_1 = \rho_1 \mathsf{P}_\ell + \mathsf{E}\mathcal{B}_1$, implying

$$\mathsf{P}_\ell = 1 - \frac{\mathsf{E}\mathcal{B}_1}{\rho_1} \ . \tag{6.2.6}$$

(Note that formula (6.2.6) is similar to formula (4.3.5).) Then we obtain

$$\lim_{t\to\infty} \frac{I_1(t)}{t} > 0 \ , \tag{6.2.7}$$

(see the discussion concerning (2.2.17)), and combining (6.2.5), (6.2.7) and relation $N_1 t = B_1(t) + I_1(t)$, we establish the strict inequality

$$\mathsf{E}\mathcal{B}_1 < N_1 . \tag{6.2.8}$$

Then by (6.2.6),

$$\rho_1 = \rho_1 \mathsf{P}_\ell + \mathsf{E}\mathcal{B}_1 < \rho_1 \mathsf{P}_\ell + N_1 ,$$

and we have proved the following *necessary stability condition of the 1st station*:

Theorem 6.7 *If the 1st station is positive recurrent, then*

$$\rho_1(1 - P_\ell) < N_1 . \tag{6.2.9}$$

Remark 6.4 In general, the quantity $\mathsf{P}_\ell = \lambda_{12}/\lambda_1$ is not explicitly available, and below we discuss how to apply stability condition (6.2.9) in practice.

Stability of the two-station system. The regeneration instants of the two-station system, with generic regeneration period T, are defined as the arrival instants of any class customers in an empty system.

Assume positive recurrence, and denote by $\mathsf{W}_2(t)$ the remaining work in station 2 at instant t, and by $I_2(t)$ the aggregated idle time of all servers in station 2 during the interval $[0, t]$. Now we may write down the following balance equation for the amount of work in the 2nd station:

$$\mathsf{W}_2(0^-) + V_2(t) + V_{12}(t) = \mathsf{W}_2(t) + N_2 t - I_2(t) , \tag{6.2.10}$$

where the amount of work $V_{12}(t)$ that has arrived in station 2 from station 1, in the time interval $[0, t]$, satisfies (6.1.14). It follows from (6.2.4) that, w.p.1,

$$\lim_{t\to\infty} \frac{V_{12}(t)}{t} = \frac{\lambda_1}{\mu_{12}} \mathsf{P}_\ell = \frac{\lambda_{12}}{\mu_{12}} , \tag{6.2.11}$$

and moreover,

$$\lim_{t\to\infty} \frac{V_2(t)}{t} = \lim_{t\to\infty} \frac{1}{t} \sum_{k=1}^{A_2(t)} S_k^{(2)} = \rho_2 , \tag{6.2.12}$$

$$\lim_{t\to\infty} \frac{I_2(t)}{t} > 0 , \tag{6.2.13}$$

where inequality (6.2.13) is a consequence of the positive recurrence assumption, see (6.2.7). Using equality (6.2.6), we obtain from (6.2.10)–(6.2.13) the following *necessary stability condition for the entire system*:

$$\rho_2 + \frac{\lambda_1}{\mu_{12}} \mathsf{P}_\ell = \rho_2 + \frac{\lambda_1 - \mu_1 \mathsf{E}\mathcal{B}_1}{\mu_{12}} < N_2 . \tag{6.2.14}$$

Also, we would like to point out that positive recurrence of the whole system implies positive recurrence of the 1st station. We now summarize the previous results in the following theorem:

Theorem 6.8 *If the 1st station is positive recurrent, then condition (6.2.9) holds. If the two-station system is positive recurrent, then both conditions (6.2.14) and (6.2.9) hold.*

Remark 6.5 It is easy to see that the analysis developed above is valid as well for the system considered in [1] with *non-preemptive* priority of class-(1, 2) customers in the 2nd station.

6.2.3 Sufficient Stability Conditions

Stability of the 1st station. Assume that the interarrival time τ and service times $S^{(j)}$ satisfy the regeneration conditions

$$\mathsf{P}(\tau > S^{(j)}) > 0\,, \; j = 1, 2\,. \tag{6.2.15}$$

In order to deduce *sufficient* stability conditions for the 1st station, we assume that $\mathsf{E}T^{(1)} = \infty$. Then we easily obtain (as in Theorem 2.2) that, for each k,

$$\mathsf{P}(Q_1(t) \geq k) \to 1, \;\; t \to \infty\,. \tag{6.2.16}$$

Denote by $J_{12}^{(i)}(t)$ the time that server i (in station 2) is occupied by class-(1, 2) customers in the interval $[0, t]$. Then it follows from (6.2.16) that, for each i,

$$\lim_{t\to\infty} \frac{\mathsf{E}J_{12}^{(i)}(t)}{t} = 1\,. \tag{6.2.17}$$

Now denote by $Z_{12}^{(i)}(t)$ the number of renewals in the process generated by the service times of class-(1, 2) customers in server i of station 2 during time interval $[0, t]$. Then, using the representation

$$\mathsf{E}A_{12}(t) = \int_0^t \sum_{i=1}^{N_2} \mathsf{E}Z_{12}^{(i)}(u)\mathsf{P}(J_{12}^{(i)}(t) \in du)\,,$$

and (6.2.17) we obtain, as in (6.1.26), that

$$\lim_{t\to\infty} \frac{1}{t}\mathsf{E}A_{12}(t) = N_2\mu_{12}\,. \tag{6.2.18}$$

Consequently, class-(1, 2) customers eventually completely capture the 2nd station, and the output from station 2 approaches a process which is the superposition of the iid *renewal processes* generated by the service times of class-(1, 2) customers realized in all N_2 servers with total service rate $N_2\mu_{12}$. In view of (6.2.16),

$$\lim_{t\to\infty}\frac{1}{t}\int_0^t\sum_{i=1}^{N_1}\mathsf{P}(W_1^{(i)}(u)>0)du=N_1,\ t\to\infty\,,$$

is the the limiting average number of busy servers in the 1st station, while

$$\lim_{t\to\infty}\frac{1}{t}\mathsf{E}I_1(t)=\lim_{t\to\infty}\frac{1}{t}\int_0^t\sum_{i=1}^{N_1}\mathsf{P}(W_1^{(i)}(u)=0)du=0\,. \qquad (6.2.19)$$

Moreover, in view of (6.1.20), and using (6.2.18), we obtain

$$\lim_{t\to\infty}\frac{1}{t}\mathsf{E}L_1(t)=\lim_{t\to\infty}\frac{1}{\mu_1}\frac{\mathsf{E}A_{12}(t)}{t}=\frac{N_2\mu_{12}}{\mu_1}\,. \qquad (6.2.20)$$

Finally, rewriting (6.2.2) as

$$\mathsf{W}_1(t)=V_1(t)+\mathsf{W}_1(0^-)-L_1(t)-N_1t+I_1(t)\ge 0\,,$$

and using (6.2.19) and (6.2.20), we find that

$$\lim_{t\to\infty}\frac{1}{t}\mathsf{E}\Big[V_1(t)+\mathsf{W}_1(0^-)-L_1(t)-N_1t+I_1(t)\Big]=\rho_1-\frac{N_2\mu_{12}}{\mu_1}-N_1\ge 0\,.$$

Consequently, the *necessary condition* for the 1st station to be *not positive recurrent* is given by the inequality

$$\lambda_1\ge N_1\mu_1+N_2\mu_{12}\,.$$

Therefore, we have proved the following sufficient positive recurrence condition of the 1st station:

Theorem 6.9 *If condition*

$$\lambda_1< N_1\mu_1+N_2\mu_{12}\,, \qquad (6.2.21)$$

is satisfied, then the (zero initial state) 1st station is positive recurrent, $\mathsf{E}T^{(1)}<\infty$.

Partial stability. Next, in the zero initial state setting, we deduce the conditions for the 1st station to remain positive recurrent (under condition (6.2.21)), while the 2nd station is *not positive recurrent*, meaning that for each k,

$$\mathsf{P}(Q_2(t) > k) \to 1 \,,\; t \to \infty \,. \tag{6.2.22}$$

It implies

$$\lim_{t\to\infty} \frac{1}{t}\mathsf{E}I_2(t) = \lim_{t\to\infty} \frac{1}{t}\int_0^t \mathsf{P}(\mathsf{W}_2(u) = 0)du = 0 \,. \tag{6.2.23}$$

Rewriting (6.2.10) then leads to

$$\mathsf{W}_2(t) = \mathsf{W}_2(0^-) + V_2(t) + V_{12}(t) - N_2 t + I_2(t) \geq 0 \,. \tag{6.2.24}$$

Also note that positive recurrence of the 1st station in particular means that $\{V_{12}(t)\}$ and $\{A_{12}(t)\}$ are positive recurrent processes with regenerative increments. Now, in order to study the asymptotic behavior of the balance equation (6.2.24), we will use convergence w.p.1 instead of convergence in mean, because the latter requires the imposition of extra moment assumptions which are difficult to formulate in terms of the given variables, see (1.2.18). However, while the time-average convergence w.p.1 holds for the positive recurrent processes $\{V_2(t)\}$, $\{V_{12}(t)\}$ and $\{A_{12}(t)\}$, we only have the convergence in mean (6.2.23) for the idle time process $\{I_2(t)\}$ at our disposal. To overcome this problem, we apply *Markov's inequality*,

$$\mathsf{E}I_2(t) \geq \varepsilon t \mathsf{P}(I_2(t) > \varepsilon t) \,,$$

where $\varepsilon > 0$ is arbitrary, implying from (6.2.23) convergence of $I_2(t)/t$ to zero *in probability*:

$$\lim_{t\to\infty} \mathsf{P}\Big(\frac{I_2(t)}{t} > \varepsilon\Big) \leq \lim_{t\to\infty} \frac{\mathsf{E}I_2(t)}{\varepsilon t} = 0 \,.$$

Hence,

$$\lim_{k\to\infty} \mathsf{P}\Big(\frac{I_2(u_k)}{u_k} > \varepsilon\Big) \to 0 \,,$$

for any (deterministic) sequence of time instants $u_k \to \infty$. Following [7] (Chap. 6), now we take a subsequence $\{z_n := u_{k_n}\} \subseteq \{u_k\}$, such that $z_n \to \infty$, $n \to \infty$, and for each n,

$$\mathsf{P}\Big(\frac{I_2(z_n)}{z_n} > \varepsilon\Big) \leq \frac{1}{n^2} \,.$$

This implies

$$\sum_{n=1}^{\infty} \mathsf{P}\Big(\frac{I_2(z_n)}{z_n} > \varepsilon\Big) < \infty \,,$$

and it follows from the Borel–Cantelli lemma [7] that only a finite w.p.1 number of events $\{I_2(z_n)/z_n > \varepsilon\}$ occur. Then, since ε has been arbitrarily chosen,

$$\frac{I_2(z_n)}{z_n} \to 0 \quad \text{w.p.1 as } n \to \infty\,. \tag{6.2.25}$$

Therefore, in the balance equation (6.2.24), we take limits w.p.1 along the subsequence $\{z_n\}$, also using the positive recurrence of the processes in the 1st station. Note that, in particular,

$$\lim_{n\to\infty} \frac{1}{z_n} V_{12}(z_n) = \frac{\lambda_1}{\mu_{12}} \mathsf{P}_\ell = \frac{\lambda_{12}}{\mu_{12}}\,, \quad \lim_{n\to\infty} \frac{1}{z_n} A_{12}(z_n) = \lambda_{12}\,, \tag{6.2.26}$$

(analogous to the limits (6.1.33), (6.2.11)). The limit P_ℓ is now obtained under the assumption (6.2.22), but this is the same as in (6.2.4), because the 1st station acts *independently* of the serving process of class-2 customers. By (6.2.12), w.p.1,

$$\lim_{n\to\infty} \frac{1}{z_n} V_2(z_n) = \rho_2\,. \tag{6.2.27}$$

Now, taking into account (6.2.23)–(6.2.27) and expression (6.2.6), we obtain the following *necessary instability condition* of the 2nd station:

$$\lim_{n\to\infty} \frac{\mathsf{W}_2(z_n)}{z_n} = \rho_2 + \frac{\lambda_1}{\mu_{12}} \mathsf{P}_\ell - N_2 = \rho_2 + \frac{\lambda_1 - \mu_1 \mathsf{E}\mathcal{B}_1}{\mu_{12}} - N_2 \geq 0\,.$$

In other words, under assumption (6.2.22), it follows that

$$\rho_2 + \frac{\lambda_1 - \mu_1 \mathsf{E}\mathcal{B}_1}{\mu_{12}} \geq N_2\,. \tag{6.2.28}$$

The inequality (6.2.28) can be rewritten as

$$\lambda_1 \geq \mu_1 \mathsf{E}\mathcal{B}_1 + N_2\mu_{12} - \rho_2\mu_{12}\,.$$

On the other hand, because $\mathsf{E}\mathcal{B}_1 < N_1$ (see (6.2.8)), we obtain

$$N_1\mu_1 + N_2\mu_{12} > \mu_1\mathsf{E}\mathcal{B}_1 + N_2\mu_{12} > \mu_1\mathsf{E}\mathcal{B}_1 + N_2\mu_{12} - \rho_2\mu_{12}\,.$$

Consequently we have hereby proved the following theorem:

Theorem 6.10 *If the the 1st station is positive recurrent and the 2nd station is not positive recurrent, then the following inequalities are satisfied*

$$\begin{aligned} \mu_1\mathsf{E}\mathcal{B}_1 + N_2\mu_{12} - \rho_2\,\mu_{12} &\leq \lambda_1 \\ &< N_1\mu_1 + N_2\mu_{12}\,. \end{aligned}$$

Stability of the whole (zero initial state) system. Assume that condition (6.2.14) holds, which is opposite to instability condition (6.2.28) of the 2nd station. Hence, $Q_2(t) \not\Rightarrow \infty$, and then

$$\inf_n \mathsf{P}(Q_2(z_n) \le C) \ge \varepsilon\,, \tag{6.2.29}$$

for some constants $C \ge 0$, $\varepsilon > 0$ and a deterministic sequence $z_n \to \infty$. On the other hand, rewriting inequality (6.2.14) as follows

$$\lambda_1 < \mu_{12}N_2 + \mu_1\mathsf{E}\mathcal{B}_1 - \rho_2\mu_{12} < \mu_{12}N_2 + \mu_1 N_1\,,$$

we observe that (6.2.21) holds, and hence, the positive recurrence of the 1st station is valid. Thus, under condition (6.2.14), the process $\{Q_1(t)\}$ is positive recurrent, *tight*, and it follows from (6.2.29) that

$$\inf_n \mathsf{P}\big(Q_2(z_n) \le C,\ Q_1(z_n) \le C_1\big) \ge \varepsilon/2\,,$$

for some constant C_1. Then in the standard way as adopted before (and also using regeneration conditions (6.2.15)), we can deduce that $\mathsf{E}T < \infty$, and hence the positive recurrence of the processes

$$\mathbf{Q}(t) := (Q_1(t),\ Q_2(t))\,,\quad \mathbf{W}(t) := (\mathsf{W}_1(t),\ \mathsf{W}_2(t))\,,\ t \ge 0\,.$$

Consequently, the previous analysis can be summarized by the following theorem:

Theorem 6.11 *The preemptive-resume two-station N-model with zero initial state is positive recurrent under assumptions (6.2.14) and (6.2.15).*

Remark 6.6 The practical use of condition (6.2.14) is limited by the presence of the unknown probability P_ℓ (or, equivalently, $\mathsf{E}\mathcal{B}_1$ or λ_{12}). In Sect. 6.3 we discuss how this problem can be resolved.

6.3 Notes

Both a parallel, as well as a serial network configuration of flexible servers may be considered, and such networks have indeed been studied on many occasions (just to mention a few, see e.g. [8–11] and the references therein for some reading on this topic). Closely related to this is the concept of *cross-trained* servers (see [12–15], and the references therein) where one station of servers is fit to handle a limited set of customers, whereas a second station has been trained to handle all types of customers that enter. Depending on the specific context, the above mentioned queueing networks can be applied to model a variety of real-life systems, such as front or back room operations in a call or service center, production systems, computer networks with rescheduling of jobs, multiserver systems with heterogeneous job types, etc.

The analysis of the cascade N-model is inspired by the stability analysis of a Markovian variant of such a two-server model by means of the Lyapunov function approach developed in [16]. The stability analysis of this model with an arbitrary number of stations has been studied by means of a fluid approach in [17]. Paper [18] (also see [19]) is devoted to an optimal allocation in a two-server system which we consider in Sect. 6.1. The authors use the *complete resource polling* (CRP) assumption $\lambda_1 > \mu_1$, or the *resource polling* (RP) assumption $\lambda_1 = \mu_1$, in order to obtain the boundary of the *maximal stability region* (see also (6.1.6))

$$\rho_2 + \frac{\lambda_1 - \mu_1}{\mu_{12}} = 1 . \qquad (6.3.1)$$

Paper [20] is devoted to optimal scheduling for the CRP scenario. Denote by ϕ_{ij} the (long-run) fraction of time that station j is allocated to class-i customers, $i, j = 1, 2$. Then the boundary of the stability region (under the CRP assumption $\lambda_1 > \mu_1$) given in [20] can be rewritten (in terms of the parameters used in our model) as

$$\lambda_1 = \mu_1\phi_{11} + \mu_{12}\phi_{12} , \quad \lambda_2 = \mu_2\phi_{22} , \qquad (6.3.2)$$

implying a maximal stability region (6.3.1) provided that

$$\phi_{22} = \rho_2 < 1, \ \phi_{11} = 1 \ \text{and} \ \phi_{12} = 1 - \rho_2 .$$

The RP assumption (3) from [21], where 'parallel servers' work as a cascade system, can be written in the same form as above.

A motivation for the model considered in Sect. 6.2 can be found in [2, 3]. The probability P_ℓ in (6.2.14) is not readily available as a closed-form expression in most cases. In part, this problem has been addressed in [1] for the case of non-preemptive priority, and by including an extra assumption. In particular, it is proved that condition (6.2.14) becomes the stability criterion, provided that P_ℓ corresponds to the *saturated system* in which the 2nd station has always class-2 customers available to serve, and for this reason the system is more easy to analyze. The analysis of the stationary regime and simulation of the two-station Markovian N-model are presented in [22].

References

1. Tezcan, T.: Stability analysis of N-model systems under a static priority rule. Queueing Syst. **73**, 235–259 (2013)
2. Whitt, W.: Blocking when service is required from several facilities simultaneously. AT&T Tech. J. **64**(8), 18071856 (1985)
3. Wong, D., Paciorek, N., Walsh, T., DiCelie, J., Young, M., Peet, B.: Concordia: An infrastructure for collaborating mobile agents. In: Proceedings Mobile Agents: First International Workshop (LNCS 1219), pp. 86–97. Berlin (1997)

4. Ahn, H.-S., Duenyas, I., Zhang, Q.R.: Optimal control of a flexible server. Adv. Appl. Prob. **36**, 139–170 (2004)
5. Morozov, E., Steyaert, B.: Stability analysis of a two-station cascade queueing network. Ann. Oper. Res. **202**, 135–160 (2013)
6. Morozov, E.: Stability of a two-pool N-model with preemptive-resume priority. In: Proceedings International Conference on Distributed Computer and Communication Networks (DCCN 2018), pp. 399–409. Moscow (2018)
7. Borovkov, A.A.: Probability Theory. Springer Science and Business Media, Berlin (2013)
8. Dai, J., Hasenbein, J., Kim, B.: Stability of join-the-shortest-queue networks. Queueing Syst. **57**, 129–145 (2007)
9. Down, D.G., Lewis, M.E.: Dynamic load balancing in parallel queueing systems: stability and optimal control. Eur. J. Oper. Res. 168, 509–519 (2006)
10. Kirkizlar, E., Andradottir, S., Ayhan, H.: Robustness of efficient server assignment policies to service time distributions in finite-buffered lines. Nav. Res. Logist. **57**(6), 563–582 (2010)
11. Tsai, Y.C., Argon, N.T.: Dynamic server assignment policies for assembly-type queues with flexible servers. Nav. Res. Logist. **55**(3), 234–251 (2008)
12. Agnihothri, S.R., Mishra, A.K., Simmons, D.E.: Workforce cross-training decisions in field service systems with two job types. J. Oper. Res. Soc. **54**(4), 410–418 (2003)
13. Ahghari, M., Balcioglu, B.: Benefits of cross-training in a skill-based routing contact center with priority queues and impatient customers. IIE Trans. **41**, 524–536 (2009)
14. Tekin, E., Hopp, W.J., Van Oyen, M.P.: Pooling strategies for call center agent cross-training. IIE Trans. **41**(6), 546–561 (2009)
15. Terekhov, D., Beck, J.C.: An extended queueing control model for facilities with front room and back room operations and mixed-skilled workers. Eur. J. Oper. Res. **198**(1), 223–231 (2009)
16. Foley, R.D., McDonald, D.R.: Large deviations of a modified Jackson network: Stability and rough asymptotics. Ann. Appl. Prob. **15**(1B), 519–541 (2005)
17. Delgado, R., Morozov, E.: Stability analysis of cascade networks via fluid models. Perf. Eval. **82**, 39–54 (2014)
18. Bell, S.L., Williams, R.J.: Dynamic scheduling of a system with two parallel servers in heavy traffic with resource pooling: Asymptotic optimality of a threshold policy. Ann. Appl. Prob. **11**(3), 608–649 (2001)
19. Bell, S.L., Williams, R.J.: Dynamic scheduling of a server system with two parallel servers: asymptotic optimality of a continuous review threshold policy in heavy traffic. In: Proceedings $38th$ Conference on Decision and Control, pp. 2255–2260. Phoenix Az. (1999)
20. Mandelbaum, A., Stolyar, A.L.: Scheduling flexible servers with convex delay costs: heavy-traffic optimality of the generalized $c\mu$-rule. Oper. Res. **52**(6), 836–855 (2004)
21. Andradottir, S., Ayhan, H., Down, G.D.: Dynamic server allocation for queueing networks with flexible servers. Oper. Res. **51**(6), 952–968 (2003)
22. Maltseva, M., Morozov, E.: Stability Analysis and Simulation of an N-Model with Two Interacting Pools. In: Proceedings The First International Conference "Stochastic Modeling and Applied Research of Technology" (SMARTY'18), pp. 77–89. Petrozavodsk (2018)

Chapter 7
Multiclass Retrial Systems with Constant Retrial Rates

In a loss queueing system, a customer facing a busy server upon arrival (a blocked customer) leaves the system forever without waiting for service. However, in many applications, the blocked customers leaving the system may retry to enter service after some random time period. In this case, it is assumed that the blocked customer waits in a virtual waiting space (an infinite *orbit*) outside the system, before retrying to access the server. These queueing systems are known as *retrial queues*.

7.1 Description of the Model

Following [1], we consider a single-server retrial system, depicted in Fig. 7.1, denoted by Σ, with K classes of *balking customers* who may or may not join the orbit queue of infinite size if the server is busy upon their arrival. Class-k customers follow a Poisson process with instants $\{t_n^{(k)}\}$ and rate λ_k, and are assumed to have iid general service times $\{S_n^{(k)}\}$ with mean $\mathsf{E}S^{(k)} = 1/\mu_k$. Define the quantities

$$\lambda = \sum_{k=1}^{K} \lambda_k \, , \quad p_k = \frac{\lambda_k}{\lambda} \, , \quad \rho_k = \frac{\lambda_k}{\mu_k} \, , \quad \rho = \sum_{k=1}^{K} \rho_k \, .$$

The aggregated arrival process is Poisson with rate λ, and any newly arriving customer belongs to class k with probability p_k and, finding the server busy upon arrival, joins the class-k orbit (also called the k-orbit in this chapter) with probability b_k, or leaves the system forever with probability $1 - b_k$. More precisely, for each class k, we consider the series of iid Bernoulli random variables $\{\beta_n^{(k)}\}$ such that $\mathsf{E}\beta^{(k)} = b_k$, and the nth class-k customer observing a busy server upon arrival joins the k-orbit if $\beta_n^{(k)} = 1$. Retrial attempts from the (non-empty) k-orbit follow a Poisson distribution with rate $\mu_0^{(k)}$, independent of the orbit size (the number of customers in the orbit) and

E. Morozov and B. Steyaert, *Stability Analysis of Regenerative Queueing Models*,
https://doi.org/10.1007/978-3-030-82438-9_7

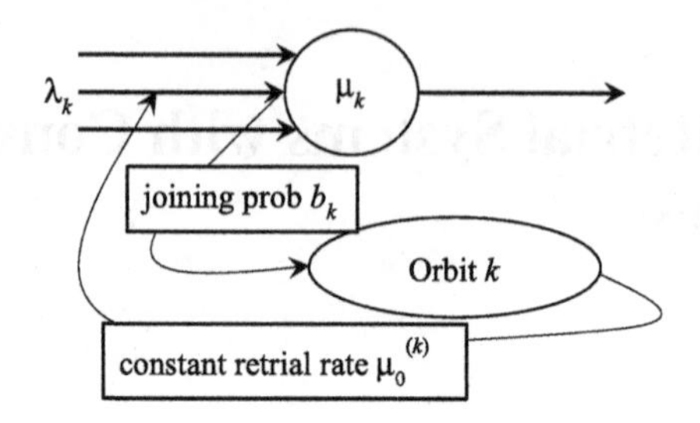

Fig. 7.1 A K-class single-server retrial system, with customer balking and constant retrial rates

the state of the server (busy or idle). It is worth mentioning that if $b_k \equiv 1$, then the present model incorporates conventional *constant retrial rates* multiclass systems.

Denote by $N_k(t)$ the number of class-k orbital customers and by $Q(t) \in \{0, 1\}$ the number of customers in the server respectively, at instant t. Let $\{t_n\}$ be the arrival instants of the *superposed* (Poisson) input process with rate λ. We consider the basic (non-Markovian) process

$$X(t) = \sum_{k=1}^{K} N_k(t) + Q(t),\ t \geq 0\,, \tag{7.1.1}$$

which describes the total number of customers in the system, and which regenerates at the instants

$$T_{n+1} = \inf_{k\geq 1}\big(t_k > T_n : X(t_k^-) = 0\big),\ \ n \geq 0\ (T_0 := 0)\,. \tag{7.1.2}$$

As always, we denote by T a generic regeneration period, and note that a new regeneration period is initiated by a class-k customer with probability p_k.

This seems a suitable occasion to attract the reader's attention to the fact that the distribution of the process $\{X(t)\}$ at a regeneration instant in this case is 'spread' over the set $\{1, \ldots, K\}$, in the sense that, with probability p_k, a new regeneration cycle, denoted by $T^{(k)}$, is initiated by a class-k customer, and its distribution in general depends on k. Therefore, the 'unconditional' generic period T can be represented as

$$T =_{st} \sum_{k=1}^{K} 1_k T^{(k)}\,,$$

where the indicator function $1_k = 1$ if a class-k customer generates a new regeneration cycle, and 0 otherwise. In this regard see also the discussion the measure φ in Sect. 2.6.

Assuming positive recurrence, we first deduce some stationary probabilities. These results have an independent interest, but are then used as well to deduce the *necessary stability conditions* of the system.

7.2 Steady-State Analysis

We define the following *stationary probabilities*:

P_0, the server is idle;
$\mathsf{P}_B = 1 - \mathsf{P}_0$, the server is busy;
$\mathsf{P}_{0,0}^{(k)}$, the server is idle and k-orbit is empty;
$\mathsf{P}_{0,b}^{(k)}$, the server is idle and k-orbit is non-empty;
$\mathsf{P}_b^{(k)}$, the server is occupied by a class-k customer.
Also, let

$$C = 1 + \rho - \sum_{k=1}^{K} \rho_k b_k \,.$$

Theorem 7.1 *If the* Σ *retrial system is positive recurrent, then*

$$P_0 = 1 - \frac{\rho}{C} \,, \tag{7.2.1}$$

and, for each $k = 1, \ldots, K$, *the following relations hold:*

$$P_b^{(k)} = \rho_k \big(1 - \frac{\rho}{C}(1 - b_k)\big) \,; \tag{7.2.2}$$

$$P_{0,\,b}^{(k)} = \frac{\lambda_k b_k}{\mu_0^{(k)}} \frac{\rho}{C} \,; \tag{7.2.3}$$

$$P_{0,\,0}^{(k)} = 1 - \Big(1 + \frac{\lambda_k b_k}{\mu_0^{(k)}}\Big) \frac{\rho}{C} \,. \tag{7.2.4}$$

Proof Define the embedded process

$$Q(t_n^{(k)-}) = Q_n^{(k)}, \; n \ge 1; \; k = 1, \ldots, K \,,$$

representing the number of customers (0 or 1) in the server at the arrival instant of the nth class-k customer. Also, for each k, we define the iid sequence of indicator functions $\{\mathsf{1}_n^{(k)}\}$, with generic element $\mathsf{1}^{(k)}$, such that $\mathsf{1}_n^{(k)} = 1$ if the nth arrival is a class-k customer, and $\mathsf{1}_n^{(k)} = 0$ otherwise. Note that, for each k, the sequence $\{\mathsf{1}_n^{(k)}\}$ is iid with $\mathsf{E}\mathsf{1}^{(k)} = p_k$, and that $\mathsf{1}(Q_n^{(k)} = 1) = Q_n^{(k)}$. Define $A_k(t)$ as the number of class-k customer arrivals in the time interval $[0,\ t]$, and let $A(t) = \sum_k A_k(t)$. Then, the amount of work $V_k(t)$ that class-k customer arrivals add to the system during $[0,\ t]$ can be represented by

$$V_k(t) = \sum_{n=1}^{A(t)} \mathsf{1}_n^{(k)} S_n^{(k)} \Big(\mathsf{1}(Q_n^{(k)} = 0) + Q_n^{(k)} \beta_n^{(k)}\Big) \,, \tag{7.2.5}$$

while the idle time of the server satisfies

$$I(t) = t - \sum_{k=1}^{K} B_k(t) = t - B(t) ,$$

where $B_k(t)$ is the amount of time devoted to *class-k* customers by the server. Denote, at instant t, by $W_k(t)$ the workload process in the k-orbit and by $S_k(t)$ the remaining service time of a class-k customer, assuming that $S_k(t) = 0$, if the server is either idle or serves a class-j customer, $j \neq k$. Then the following balance equation holds:

$$W_k(0^-) + V_k(t) = S_k(t) + B_k(t) + W_k(t),\ t \geq 0;\ k = 1, \dots, K . \tag{7.2.6}$$

In view of the positive recurrence of the involved processes and because the input is Poisson, then the limit $Q_n^{(k)} \Rightarrow Q^{(k)},\ n \to \infty$, exists and represents the stationary state of the server observed by a newly arriving class-k customer. Moreover, the stationary probabilities

$$\mathsf{P}_b^{(k)} = \lim_{t\to\infty} \frac{B_k(t)}{t} = \lim_{t\to\infty} \mathsf{P}(S_k(t) > 0) = \mathsf{P}(\mathbb{S}_k > 0) ,\ \ k = 1, \dots, K , \tag{7.2.7}$$

exist, where $\mathbb{S}_k$ is the limit, $S_k(t) \Rightarrow \mathbb{S}_k$. Moreover w.p.1 (see (2.3.22)),

$$\lim_{t\to\infty} \frac{S_k(t)}{t} = \lim_{t\to\infty} \frac{W_k(t)}{t} = 0 ,\ \ k = 1, \dots, K . \tag{7.2.8}$$

Note that $\{V_k(t)\}$ is a process with regenerative increments and mean (cycle) increment upper bounded by $\mathsf{E}T < \infty$. Because $A(t)/t \to \lambda$, then we obtain from (7.2.5) that, w.p.1,

$$\begin{aligned}\lim_{t\to\infty} \frac{V_k(t)}{t} &= \lim_{t\to\infty} \frac{A(t)}{t} \frac{1}{A(t)} \sum_{n=1}^{A(t)} 1_n^{(k)} S_n^{(k)} \big(1(Q_n^{(k)} = 0) + Q_n^{(k)} \beta_n^{(k)}\big) \\ &= \lambda \mathsf{E}\Big[1^{(k)} S^{(k)} \Big(1(Q^{(k)} = 0) + Q^{(k)} \beta^{(k)}\Big)\Big] \\ &= \lambda p_k \mathsf{E} S^{(k)} (\mathsf{P}_0 + \mathsf{P}_B b_k) = \rho_k (\mathsf{P}_0 + \mathsf{P}_B b_k) ,\end{aligned} \tag{7.2.9}$$

where we have applied (2.1.7) to find $\lim_t 1/A(t) \sum_1^{A(t)}$ and relied on the independence of the random variables $S^{(k)}$, $\beta^{(k)}$ and the indicator functions. In view of the PASTA property, the probability P_B (P_0) is also the limiting fraction of *time* that the server is busy (idle), and these limits do not depend on the customer class. Then it is easy to conclude from (7.2.6)–(7.2.9) that

$$\mathsf{P}_b^{(k)} = \rho_k (\mathsf{P}_0 + \mathsf{P}_B b_k),\ \ k = 1, \dots, K . \tag{7.2.10}$$

Denote the aggregated work $V(t) = \sum_{k=1}^{K} V_k(t)$. Now we sum equations (7.2.6) over all classes and obtain, from (7.2.9) and (7.2.10), that

$$\lim_{t\to\infty} \frac{V(t)}{t} = \sum_{k=1}^{K} \rho_k(\mathsf{P}_0 + \mathsf{P}_B b_k)\,, \tag{7.2.11}$$

while by using (7.2.8),

$$\lim_{t\to\infty} \frac{1}{t}\Big(\sum_{k=1}^{K}(S_k(t) + W_k(t)) + t - I(t)\Big) = 1 - \mathsf{P}_0\,. \tag{7.2.12}$$

The limit

$$\lim_{t\to\infty} \frac{I(t)}{t} = \lim_{t\to\infty} \mathsf{P}(Q(t) = 0) = \mathsf{P}(Q = 0) = \mathsf{P}_0\,,$$

represents the stationary *idle probability* of the server, with Q the stationary number of customers in the server. As before, due to the PASTA property,

$$\begin{aligned} \mathsf{P}_0 &= \mathsf{P}(Q = 0) = \mathsf{P}(Q^{(k)} = 0)\,, \\ \mathsf{P}_B &= \mathsf{P}(Q = 1) = \mathsf{E}Q = \mathsf{E}Q^{(k)},\ k = 1, \ldots, K\,. \end{aligned} \tag{7.2.13}$$

It now becomes clear that (7.2.1) follows from (7.2.11) and (7.2.12), while (7.2.2) follows from (7.2.1) and (7.2.10).

It is much more challenging to derive an expression for $\mathsf{P}^{(k)}_{0,\,b}$. Fix some k and denote by $D_k(t)$ the number of class-k retrial customers that depart the k-orbit within the time interval $[0,\ t]$. Also, let $\widehat{\mathbb{D}}_k = \{\widehat{D}_k(t),\ t \ge 0\}$ represent a Poisson process with parameter $\mu_0^{(k)}$, then the probability of having i events in the time interval $[0,\ t]$ equals

$$\mathsf{P}(\widehat{D}_k(t) = i) = e^{-\mu_0^{(k)} t}\, \frac{[\mu_0^{(k)} t]^i}{i!},\quad i \ge 0\,. \tag{7.2.14}$$

Denote by $\{u_n^{(k)},\ n \ge 1\}$ the renewal instants of the process $\widehat{\mathbb{D}}_k$. Now we use a coupling argument to relate the process of real departures to the Poisson process $\widehat{\mathbb{D}}_k$. We sample the process $\widehat{\mathbb{D}}_k$ until the first class-k customer joins the orbit. At this instant, we resample the remaining renewal (exponential) time in the process $\widehat{\mathbb{D}}_k$ and then treat the subsequent renewal intervals in the process $\widehat{\mathbb{D}}_k$ as the intervals between the attempts from the k-orbit, until a class-k orbital customer leaves the orbit empty. From this instant on, we then continue to sample the process $\widehat{\mathbb{D}}_k$ until the next *real* customer joins the k-orbit at some instant v_k (while the process of real attempts originating from orbit k remains 'frozen' until instant v_k). At instant v_k, we resample the remaining renewal time in the process $\widehat{\mathbb{D}}_k$ and, as above, interpret the subsequent renewals in the process $\widehat{\mathbb{D}}_k$ as the attempts from orbit k, until the orbit becomes empty again, etc. The modified renewal process (with resampling) is stochastically equivalent to originally defined process $\widehat{\mathbb{D}}_k$. In what follows, we therefore keep the same notation $\widehat{\mathbb{D}}_k$ for this modified process, and $\{u_n^{(k)},\ n \geq 1\}$ for its renewal instants.

By construction, the instants of attempts of the (top) orbital class-k customers constitute a *subsequence* of the renewal instants of the process $\widehat{\mathbb{D}}_k$ with resampling. Next, for each k, let

$$\mathcal{Q}(u_n^{(k)-}) = \mathcal{Q}_n^{(k)}, \quad N_k(u_n^{(k)-}) = N_n^{(k)}, \quad n \geq 1 .$$

(We adopt the notation $\mathcal{Q}$ to distinguish from the process $\{Q_n^{(k)}\}$ embedded at the *arrival* instants $\{t_n\}$.) It follows from the above construction that the equality

$$\mathsf{1}(\mathcal{Q}_n^{(k)} = 0,\ N_n^{(k)} > 0) = 1 ,$$

implies that the nth renewal instant of the renewal process $\widehat{\mathbb{D}}_k$ is a *successful attempt* of a class-k customer to enter the server. Consequently, the number of customers which leave the k-orbit in the time interval $[0,\ t]$ satisfies

$$D_k(t) = \sum_{n=1}^{\widehat{D}_k(t)} \mathsf{1}(\mathcal{Q}_n^{(k)} = 0,\ N_n^{(k)} > 0),\ t \geq 0;\ k = 1, \ldots, K . \quad (7.2.15)$$

Since the orbit is not idle at the instant of each such an attempt, a key observation is that the number of successful attempts from orbit k is equal to the number of events generated by the Poisson process $\widehat{\mathbb{D}}_k$ which *simultaneously observe an empty server and a busy k-orbit*. Due to the positive recurrence of the involved processes, the limit

$$\lim_{t\to\infty} \frac{1}{\widehat{D}_k(t)} \sum_{n=1}^{\widehat{D}_k(t)} \mathsf{1}(\mathcal{Q}_n^{(k)} = 0,\ N_n^{(k)} > 0) = \mathsf{P}_{0,b}^{(k)} , \quad (7.2.16)$$

exists for each k. We again note that, by invoking the PASTA property, the probability $\mathsf{P}_{0,b}^{(k)}$ is also the limiting fraction of the time during which the server is idle and the kth orbit is busy. Moreover, by the SLLN (1.2.10)

$$\lim_{t\to\infty} \frac{\widehat{D}_k(t)}{t} = \mu_0^{(k)} , \quad (7.2.17)$$

and therefore, Eqs. (7.2.15)–(7.2.17) result in

$$\lim_{t\to\infty} \frac{D_k(t)}{t} = \mu_0^{(k)} \mathsf{P}_{0,b}^{(k)} , \quad k = 1, \ldots, K . \quad (7.2.18)$$

Denote by $J_k(t)$ the number of class-k customers that join orbit k in the time interval $[0,\ t]$, then

$$J_k(t) = \sum_{n=1}^{A(t)} \mathcal{Q}_n^{(k)} 1_n^{(k)} \beta_n^{(k)}, \quad k = 1, \ldots, K .$$

Note that the cycle increment of the positive recurrent (integer-valued) process $\{J_k(t)\}$ with regenerative increments equals its maximum over a cycle and is upper bounded by the cycle length θ. Because $\mathsf{E}\theta = \lambda \mathsf{E}T < \infty$, it then follows from (7.2.13) and (2.1.7) that

$$\begin{aligned}\lim_{t\to\infty} \frac{J_k(t)}{t} &= \lim_{t\to\infty} \frac{A(t)}{t} \frac{1}{A(t)} \sum_{n=1}^{A(t)} \mathcal{Q}_n^{(k)} 1_n^{(k)} \beta_n^{(k)} \\ &= \lambda \mathsf{E}\big[\mathcal{Q}^{(k)} 1^{(k)} \beta^{(k)}\big] = \lambda_k b_k \mathsf{P}_B, \quad k = 1, \ldots, K , \qquad (7.2.19)\end{aligned}$$

where we have relied on the independence of the random variables $\beta^{(k)}$, $1^{(k)}$ and $\mathcal{Q}^{(k)}$, which is equal to the weak limit $\mathcal{Q}_n^{(k)} \Rightarrow \mathcal{Q}^{(k)}$, $n \to \infty$. Moreover, because both $\mathcal{Q}_n^{(k)}$ and $Q_n^{(k)}$ are observed at the sequence of time instants generated by Poisson processes, we can apply the PASTA property, which allows us to (stochastically) equate the limits $\mathcal{Q}^{(k)} =_{st} Q^{(k)}$, implying (also see (7.2.13))

$$\mathsf{P}(\mathcal{Q}^{(k)} = 1) = \mathsf{P}(Q^{(k)} = 1) = \mathsf{P}_B .$$

Obviously, for each k, the balance equation

$$N_k(0^-) + J_k(t) = N_k(t) + D_k(t) , \qquad (7.2.20)$$

holds as well, where, due to the positive recurrence, $N_k(t) = o(t)$, and then (7.2.3) follows from (7.2.18)–(7.2.20) by taking the appropriate limit of both sides in (7.2.20). Also, for each k, we can write down the relation

$$\mathsf{P}_0 = \mathsf{P}_{0,b}^{(k)} + \mathsf{P}_{0,0}^{(k)} , \qquad (7.2.21)$$

and consequently, (7.2.4) results from (7.2.1)–(7.2.3). □

7.3 Necessary Stability Conditions

Based on the explicit expression (7.2.3) for the stationary probability $\mathsf{P}_{0,b}^{(k)}$, we arrive at the following necessary stability conditions:

Theorem 7.2 *If system Σ is positive recurrent, then*

$$\lambda_k b_k \frac{\rho}{C} < \mu_0^{(k)} (1 - \frac{\rho}{C}) , \; k = 1, \ldots, K . \qquad (7.3.1)$$

Proof The proof is straightforward. We rewrite expression (7.2.3) as

$$\mu_0^{(k)}\mathsf{P}_{0,\,b}^{(k)} = \lambda_k b_k \frac{\rho}{C}\,, \quad k = 1, \ldots, K\,, \tag{7.3.2}$$

and note that $\mathsf{P}_{0,\,b}^{(k)} = 0$ if $b_k = 0$. Denote by τ the generic interarrival time of the merged *Poisson* input process, and let $I_0^{(k)}$ be the time period during which the server and the k-orbit are *simultaneously idle within the cycle*. Note that, with probability

$$p_k\mathsf{P}(\tau > S^{(k)} + \delta) =: \varepsilon > 0\,, \tag{7.3.3}$$

each new regeneration cycle contains only one class-k customer and the idle time of the server (and all orbits simultaneously) is no less than (an arbitrary) $\delta > 0$. This in turn yields

$$\mathsf{P}_{0,0}^{(k)} = \lim_{t\to\infty}\frac{1}{t}\int_0^t 1(Q(u) = 0,\ N_k(u) = 0)du = \frac{\mathsf{E}I_0^{(k)}}{\mathsf{E}T} > 0\,,\ k = 1, \ldots, K\,.$$

We then obtain from (7.2.1) and (7.2.21) that

$$1 - \frac{\rho}{C} = \mathsf{P}_0 = \mathsf{P}_{0,b}^{(k)} + \mathsf{P}_{0,0}^{(k)} > \mathsf{P}_{0,b}^{(k)}\,.$$

Consequently, (7.3.1) follows from (7.3.2). □

Problem 7.1 Using the assumption $\mathsf{E}S^{(k)} < \infty$, explain (7.3.3) and then show that $\mathsf{E}I_0^{(k)} \ge \delta\varepsilon$.

Remark 7.1 In the retrial system with *no balking*, $b_k \equiv 1$, implying $C = 1$, conditions (7.3.1) become

$$\lambda_k\rho = \lambda_k\mathsf{P}_B < \mu_0^{(k)}(1 - \mathsf{P}_B)\,,\ k = 1, \ldots, K\,,$$

and coincide with those found in [2] for this system. If $b_k \equiv 0$, then we obtain a loss system in which $C = 1 + \rho$, and it follows from (7.2.1) and (7.2.2) that

$$\mathsf{P}_B = \frac{\rho}{1+\rho}\,, \quad \mathsf{P}_b^{(k)} = \frac{\rho_k}{1+\rho}\,, \quad k = 1, \ldots, K\,. \tag{7.3.4}$$

7.4 Sufficient Condition: The Buffered Dominating System

In this section, we establish a sufficient stability condition for a retrial system Σ, in which there is no balking, i.e., $b_k \equiv 1$.

In the proof of Theorem 7.3 below, we heavily rely on the comparison between the retrial system Σ and the classical multiclass system (without orbits/retrials) with an infinite buffer for the waiting customers of all classes. The basic difference between the two systems is that, in the retrial system, unlike the buffered system, an idle time of the server occurs after each service completion. Recall that $\rho = \sum_{k=1}^{K} \rho_k$.

Theorem 7.3 *If* $b_k \equiv 1$ *and the condition*

$$\rho + \max_{1\le k\le K} \frac{\lambda}{\lambda + \mu_0^{(k)}} < 1\,, \tag{7.4.1}$$

holds, then $\mathsf{E}T < \infty$.

Proof Note that the service time of the nth arrival can be written as

$$S_n = \sum_{k=1}^{K} \mathsf{1}_n^{(k)} S_n^{(k)},\ n \ge 1\,. \tag{7.4.2}$$

Also note that, in order to compose the service process, we select only those elements $S_n^{(k)}$ that correspond to $\mathsf{1}_n^{(k)} = 1$; observe that $\mathsf{1}_n^{(k)}$ and $S_n^{(k)}$ are independent random variables. In the retrial model, an idle time of the server occurs after each departure, and the remaining time to the next customer arrival instant is distributed as interarrival time τ. We denote by $\{\xi_n^{(k)},\ n \ge 1\}$ the sequence of iid exponential random variables with parameter $\mu_0^{(k)}$ corresponding to the retrial times from orbit k. Then, provided orbit k is non-empty, the idle time of the server after each departure is upper bounded by the random variable $\min(\tau,\ \xi^{(k)})$ with mean

$$\mathsf{E}\min(\tau,\ \xi^{(k)}) = \frac{1}{\lambda + \mu_0^{(k)}}\,.$$

Denoting by ζ_n the idle time of the server *after the nth departure*, we observe that, if at least one orbit is non-empty, then for each $n \ge 1$,

$$\mathsf{E}\zeta_n \le \max_{1\le k\le K} \frac{1}{\lambda + \mu_0^{(k)}} = \frac{1}{\min_k(\lambda + \mu_0^{(k)})} =: r_0\,. \tag{7.4.3}$$

(If all orbits are empty, then $\mathsf{E}\zeta_n = 1/\lambda$ is the value of mean idle time preceding a regeneration instant.)

Now we construct a single-server *dominating buffered system* $\widehat{\Sigma}$, with FIFO service discipline, and the same input as in the original retrial system Σ. The only difference is that each customer in $\widehat{\Sigma}$ occupies the server, besides its basic service time, for an *extra exponential time* ξ_o with mean $\mathsf{E}\xi_o = r_0$. This implies that the generic service time $\widehat{S}$ in the system $\widehat{\Sigma}$ is distributed as $S + \xi_o$. (From now on, the corresponding variables in $\widehat{\Sigma}$ are equipped with the 'hat' superscript.) Then, using a delicate coupling construction as in [1], we prove the following intuitive result:

at each instant t, the number of customers $E(t)$ and $\widehat{E}(t)$ that have entered server within the time interval $[0,\ t]$ in the systems Σ and $\widehat{\Sigma}$ respectively, are ordered as $E(t) \geq \widehat{E}(t)$. Then the busy times of the server in both systems are ordered as $B(t) \geq \widehat{B}(t)$. Indeed, it is shown in [1] that we may assign the *same service time* to the nth customer entering the server in both systems, $n \geq 1$. This may alter the FIFO order in the buffered system, but it does not change the distribution of the remaining work. Since the amount of work that has arrived in both systems in the interval $[0,\ t]$ is the same due to the coupling approach, we may write

$$V(t) = W(t) + B(t) = \widehat{V}(t) = \widehat{W}(t) + \widehat{B}(t)\ ,$$

and then it follows that

$$\widehat{W}(t) \geq W(t)\ ,\ \ t \geq 0\ . \tag{7.4.4}$$

At the same time, the system $\widehat{\Sigma}$ is positive recurrent if the following well-known negative drift condition

$$\mathsf{E}\widehat{S} = \mathsf{E}(S + \xi_o) = \mathsf{E}S + r_0 < \mathsf{E}\tau\ ,$$

is satisfied, which in view of (7.4.2) and (7.4.3), has the following form:

$$\mathsf{E}\widehat{S} = \sum_{k=1}^{K} \frac{p_k}{\mu_k} + \frac{1}{\min_k(\lambda + \mu_0^{(k)})} < \frac{1}{\lambda}\ . \tag{7.4.5}$$

With $\lambda p_k = \lambda_k$, it is easy to check that (7.4.5) coincides with (7.4.1). Since the workload process $\{\widehat{W}(t)\}$ is positive recurrent, then the dominance inequality (7.4.4) implies that the statement of Theorem 7.3 is hereby proved. □

Remark 7.2 It is easy to deduce from the proof that the total number of customers in both systems are ordered as $\widehat{X}(t) \geq X(t),\ t \geq 0$, as well.

7.5 Sufficient Condition: A Loss Dominating System

In this section, we develop an alternative approach for the stability analysis of the *multiserver* multiclass retrial system Σ (with a *finite buffer*) without balking ($b_k \equiv 1$), by considering a special class of service time distributions. This approach, using a dominating *loss system* $\widehat{\Sigma}$ *with saturated orbits*, has been proposed in [3] for the single-class system, and then extended to a multiclass retrial system in [2].

As in the original system, the loss system accepts *primary arrivals* generated by K independent Poisson processes with rates $\{\lambda_k\}$ and total rate $\lambda = \sum_k \lambda_k$. In this loss system, retrial attempts from k-orbit occur according to a *Poisson process* with rate $\mu_0^{(k)}$ (we call this process the $\mu_0^{(k)}$*-input*), $k = 1, \ldots, K$. It is therefore convenient to envision that the $\mu_0^{(k)}$-retrial input is generated by an infinitely loaded virtual k-orbit. Moreover, we assume that the rejected class-k customers are *coloured* and join the *front* of this virtual orbit k. Consequently, if the class-k coloured customers are exhausted, then the *uncoloured customers* continue to follow the same Poisson input process with rate $\mu_0^{(k)}$.

Denote by $\{t_n\}$, the arrival instants of the aggregated primary Poisson arrival stream to the system $\widehat{\Sigma}$, and we will refer to the servers with buffer as the *primary system*. Let $\widehat{Q}(t)$ denote the number of customers in the primary system of $\widehat{\Sigma}$ at instant t. By construction, the process $\{\widehat{Q}(t)\}$ regenerates at the instants

$$Z_0 := 0\,,\ Z_{n+1} = \min_k(t_k > Z_n :\ \widehat{Q}(t_k^-) = 0)\,,\ n \geq 0\,, \tag{7.5.1}$$

which occur when a new primary customer observes an *empty primary system*. Clearly, a new regeneration period is then initiated by a class-k customer with probability

$$q_k := \frac{\lambda_k + \mu_0^{(k)}}{\sum_{j=1}^K(\lambda_j + \mu_0^{(j)})}\,,\ k = 1, \ldots, K\,.$$

Evidently, we can treat $\widehat{\Sigma}$ as *a single-class system* with (generic) service time $\widehat{S}$, defined by analogy with (7.4.2), and where each new arrival has service time $S^{(k)}$ with probability q_k. Consequently, the average service time of a customer satisfies $\mathsf{E}\widehat{S} = \sum_k q_k/\mu_k$. In addition, we define the quantities

$$\widehat{\rho}_k = \frac{\mu_0^{(k)}}{\mu_k}\,,\quad \rho_k = \frac{\lambda_k}{\mu_k}\,,\ k = 1, \ldots, K\,.$$

Because the total input rate is given by

$$\lambda = \sum_{k=1}^K(\lambda_k + \mu_0^{(k)})\,,$$

then it is easy to calculate that the system $\widehat{\Sigma}$ has the following traffic intensity:

$$\widehat{\rho} = \lambda\mathsf{E}\widehat{S} = \sum_{k=1}^K \frac{\lambda_k + \mu_0^{(k)}}{\mu_k} = \sum_{k=1}^K(\widehat{\rho}_k + \rho_k)\,. \tag{7.5.2}$$

Let $\widehat{B}(t)$ be the time, within the interval $[0,\, t]$, when the primary system (in $\widehat{\Sigma}$) is fully busy (and therefore cannot accept new customers for service). Since the primary system is finite, then the cumulative process $\{\widehat{B}(t)\}$ is positive recurrent, (with regenerations (7.5.1)) and the stationary probability $\widehat{\mathsf{P}}_B$ that the *primary system* is full is given by the limit

$$\lim_{t\to\infty}\frac{\widehat{B}(t)}{t}=\widehat{\mathsf{P}}_B\,. \tag{7.5.3}$$

Let us also define the main queueing process describing the loss system $\widehat{\Sigma}$. We set the random variable $\widehat{J}(t)=0$ if the primary system is *empty* at instant t, and $\widehat{J}(t)=1$, otherwise. Also, denote by $\widehat{N}_k(t)$ the number of *coloured* customers in orbit k at instant t, and introduce the process

$$\widehat{\mathbf{N}}(t)=\{\widehat{J}(t),\widehat{N}_1(t),\ldots,\widehat{N}_K(t)\}\,,\ t\geq 0\,,$$

which regenerates at the instants

$$\widehat{T}_{n+1}=\min(t_k>\widehat{T}_n:\widehat{\mathbf{N}}(t_k^-)=\mathbf{0}),\ \ n\geq 0\ \ (\widehat{T}_0:=0)\,,$$

when a primary customer observes an empty primary system with no coloured customers in all orbits.

We now arrive at the following theorem:

Theorem 7.4 *Assume that conditions*

$$(\lambda_k+\mu_0^{(k)})\widehat{\mathsf{P}}_B<\mu_0^{(k)},\ k=1,\ldots,K\,, \tag{7.5.4}$$

are satisfied, where $\widehat{\mathsf{P}}_B$ is defined in (7.5.3). Then the process $\{\widehat{\mathbf{N}}(t),\ t\geq 0\}$ is positive recurrent for any initial state.

Proof We summarize the main steps of the proof given in [2]. The key observation is that, *for any k*, the k-orbit size process $\{\widehat{N}_k(t)\}$ of coloured customers regenerates at those instants when a class-k arrival sees both the primary system empty and no coloured customers in orbit k, since the process $\{\widehat{N}_k(t)\}$ does not depend on the type (coloured or uncoloured) of class-j customers with $j\neq k$. At this point we also rely on the exponentiality of all interarrival times. Then we show that, if condition (7.5.4) holds for some k, the number of coloured customers in orbit k is a positive recurrent regenerative process, and therefore is tight, *irrespective* of whether or not the condition (7.5.4) is satisfied for other orbits $j\neq k$. Moreover, we rely on the insensitivity of the busy time process to the customer type (coloured/uncoloured). Consequently, each of the coloured orbit sizes is a tight process, and therefore the total number of coloured customers is a tight process as well. Finally, we show that this tightness implies positive recurrence, where a key role is played by the exponentiality of both the Poisson primary input processes and the retrial attempts. □

To formulate sufficient stability conditions for the original system Σ, we need the following definition [4]. Two non-negative absolutely continuous random variables ξ_i with distribution functions F_i, densities f_i, and *failure rates*

$$r_i(x)=\frac{f_i(x)}{1-F_i(x)}\,,$$

(defined for all x for which $F_i(x) < 1$) satisfy the *failure rate ordering* $\xi_1 \leq_r \xi_2$ if

$$\inf_{x\geq 0} r_1(x) \geq \sup_{x\geq 0} r_2(x) . \tag{7.5.5}$$

Define the indicator function $J(t) = 0$ if the primary system in Σ is empty at instant t, and $J(t) = 1$, otherwise. Also, let $N_k(t)$ be the kth orbit size in the system Σ at instant t, and define the process

$$\mathbf{N}(t) = \Big(J(t),\ N_1(t),\ \ldots, N_K(t)\Big) ,\ \ t \geq 0 ,$$

with regeneration instants

$$T_{n+1} = \min_k (t_k > T_n : \mathbf{N}(t_k^-) = \mathbf{0}) ,\ \ n \geq 0 \ (T_0 := 0) ,$$

where, as before, $\{t_n\}$ is the sequence of time instants generated by the primary merged Poisson input process with rate $\lambda = \sum_k \lambda_k$. Denote by $\exp(\mu)$ an exponential random variable with parameter μ.

Theorem 7.5 *The process* $\{\mathbf{N}(t)\}$ *is positive recurrent regenerative for any initial state under conditions (7.5.4), where the probability* $\widehat{\mathsf{P}}_B$ *is defined for the system* $\widehat{\Sigma}$ *with exponential service times with parameter* $\min \widehat{\mu}_k$ *such that*

$$S^{(k)} \leq_r \exp(\min \widehat{\mu}_k) ,\ \ k = 1, \ldots, K . \tag{7.5.6}$$

In the proof in [2] we show that, under condition (7.5.6), the service time distributions satisfy the monotonicity property (1) from [3], implying that the positive recurrent process $\{\widehat{\mathbf{N}}(t)\}$ dominates the orbit-size process $\{\mathbf{N}(t)\}$ in the system Σ.

Remark 7.3 We let F_k represent the distribution function of the (generic) service time $S^{(k)}$ and denote by f_k its density. Then, according to definition (7.5.5), condition (7.5.6) can be written as

$$\inf_{x\geq 0} \frac{f_k(x)}{1 - F_k(x)} \geq \min \widehat{\mu}_k ,\ \ k = 1, \ldots, K .$$

It is worth mentioning that the structure of the primary system (such as the number of servers, the buffer size) affects the stability conditions (7.5.4) via the busy probability $\widehat{\mathsf{P}}_B$. The practical importance of Theorem 7.4 clearly depends on the possibility to determine $\widehat{\mathsf{P}}_B$. In Sect. 7.6 below we give explicit expressions for this probability for some special cases of queueing systems.

7.5.1 Symmetrical Orbits

Now we show that the sufficient stability conditions (7.5.4) indeed coincide with the necessary conditions in a *bufferless single-server* retrial system, with a *symmetrical* parameter setting (when all classes have identical parameters) and without balking. We denote by

$$\rho_k \equiv \rho_s, \quad \frac{\mu_0^{(k)}}{\mu_k} \equiv \widehat{\rho}_s .$$

Then all K necessary stability conditions (7.3.1) are equivalent to the single inequality

$$(\widehat{\rho}_s + \rho_s) K \rho_s < \widehat{\rho}_s . \tag{7.5.7}$$

On the other hand, in this case the dominating system becomes a loss system with the traffic intensity

$$\widehat{\rho} = K(\rho_s + \widehat{\rho}_s) ,$$

and consequently the stationary loss probability satisfies (see also (7.3.4))

$$\widehat{\mathsf{P}}_B = \frac{\widehat{\rho}}{1+\widehat{\rho}} = \frac{K(\rho_s + \widehat{\rho}_s)}{1 + K(\rho_s + \widehat{\rho}_s)} .$$

Substituting this expression in the sufficient stability conditions (7.5.4) yields a single inequality

$$(\widehat{\rho}_s + \rho_s) \frac{K(\widehat{\rho}_s + \rho_s)}{1 + K(\widehat{\rho}_s + \rho_s)} < \widehat{\rho}_s . \tag{7.5.8}$$

It is easy to check that conditions (7.5.7) and (7.5.8) coincide, and yield the stability criterion for the case of symmetrical orbits.

7.5.2 Local Stability

Following [2], we now formulate (without proof) the conditions implying a *local stability* of the single-server system Σ. Recall that $\rho = \sum_{k=1}^{K} \rho_k$ and $B(t)$ represents the time during which the primary system of Σ is fully busy within the interval $[0,\ t]$.

Theorem 7.6 *Consider the zero initial state system Σ. Assume that*

$$(\lambda_j + \mu_0^{(j)})\rho \geq \mu_0^{(j)} ,$$

is satisfied for some orbit j, and that

$$(\lambda_k + \mu_0^{(k)})\widehat{\mathsf{P}}_B < \mu_0^{(k)}, \ \ k \neq j; \ k = 1, \ldots, K \ . \tag{7.5.9}$$

Then the orbit size $N_j(t) \Rightarrow \infty$, while the orbit sizes $\{N_k(t),\ t \geq 0\}$, $k \neq j$, are tight. Moreover, if

$$(\lambda_j + \mu_0^{(j)})\rho > \mu_0^{(j)},$$

then the 'busy probability becomes' (of the primary system in Σ)

$$\mathsf{P}_B := \lim_{t\to\infty} \frac{\mathsf{E}B(t)}{t} = \frac{\lambda_j + \mu_0^{(j)} + \mu_j \sum_{k\neq j} \rho_k}{\lambda_j + \mu_0^{(j)} + \mu_j}, \tag{7.5.10}$$

and $\mathsf{P}_B = \rho$, if $(\lambda_j + \mu_0^{(j)})\rho = \mu_0^{(j)}$.

To illustrate the results contained in Theorem 7.6, we consider a two-class *single-server bufferless* retrial system in which class-2 orbit satisfies assumption

$$(\lambda_2 + \mu_0^{(2)})\rho \geq \mu_0^{(2)} , \tag{7.5.11}$$

(i.e., orbit 2 is 'unstable') while assumption (7.5.9) takes the form

$$(\lambda_1 + \mu_0^{(1)})\widehat{\mathsf{P}}_B < \mu_0^{(1)} , \tag{7.5.12}$$

(implying that orbit 1 is 'stable'). In the dominating loss system $\widehat{\Sigma}$, the total input rate is equal to

$$\lambda_1 + \lambda_2 + \mu_0^{(1)} + \mu_0^{(2)} =: \widehat{\lambda} \ .$$

Because in the loss system $\widehat{\Sigma}$, class-i customers have an aggregated input rate of $\lambda_i + \mu_0^{(i)}$, then a new arrival is a class-i one (and has service rate μ_i) with probability

$$p_i := \frac{\lambda_i + \mu_0^{(i)}}{\widehat{\lambda}}, \quad i = 1, 2 ,$$

and the (generic) service time can be represented as $\widehat{S} =_{st} \mathsf{1}^{(1)} S^{(1)} + \mathsf{1}^{(2)} S^{(2)}$ with $\mathsf{E}\mathsf{1}^{(i)} = p_i$. Note that in this loss system, the regeneration cycle contains the busy period, which is equal to the service time, and the following exponential (with rate $\widehat{\lambda}$) idle period. Then, denoting by $\widehat{S}(t)$ the remaining service time in the system $\widehat{\Sigma}$ at instant t, we obtain, by using a regenerative argument, that the loss probability becomes

$$\widehat{\mathsf{P}}_B = \lim_{t\to\infty} \frac{1}{t} \int_0^t \mathsf{1}(\widehat{S}(u) > 0)du = \frac{\mathsf{E}\widehat{S}}{\mathsf{E}\widehat{S} + 1/\widehat{\lambda}} = \frac{\rho + \widehat{\rho}_1 + \widehat{\rho}_2}{\rho + \widehat{\rho}_1 + \widehat{\rho}_2 + 1}, \tag{7.5.13}$$

with $\rho = \rho_1 + \rho_2$. In this case, the busy probability P_B in the original system Σ, given by expression (7.5.10), becomes, after some simple algebra,

$$\mathsf{P}_B = \frac{\rho + \widehat{\rho}_2}{1 + \rho_2 + \widehat{\rho}_2} . \tag{7.5.14}$$

Remark 7.4 Conditions (7.5.11) and (7.5.12) imply (see Problem 7.3 below) that $\mathsf{P}_B \le \rho$, and that $\mathsf{P}_B < \rho$ if and only if inequality (7.5.11) is strict. This interesting result has the following intuitive explanation. If

$$(\lambda_2 + \mu_0^{(2)})\rho > \mu_0^{(2)} ,$$

then a *non-negligible* fraction of class-2 customers, joining an infinitely increasing orbit 2 can not enter the server in a finite time and in fact 'disappears' from the system. As a result, the limiting fraction of the 'processed' work becomes less than ρ, the arrived work per unit of time. However, if the equality $(\lambda_2 + \mu_0^{(2)})\rho = \mu_0^{(2)}$ holds, then the fraction of 'disappearing' class-2 customers becomes negligible, implying $\mathsf{P}_B = \rho$.

Remark 7.5 As a final point of this section, we explain why the probability P_B satisfies expression (7.5.14). Since the queueing process in the $1st$ orbit is tight, the probability P_B must include all arrived work ρ_1 of class-1 customers per unit of time, which is also the (limiting) fraction of time during which the server is occupied by class-1 customers. The remaining fraction of time, $1 - \rho_1$, is devoted to serving class-2 customers when the server is working as a loss system with input rate $\lambda_2 + \mu_0^{(2)}$ and service rate μ_2, in which case the loss probability equals

$$\frac{(\lambda_2 + \mu_0^{(2)})/\mu_2}{1 + (\lambda_2 + \mu_0^{(2)})/\mu_2} = \frac{\rho_2 + \widehat{\rho}_2}{1 + \rho_2 + \widehat{\rho}_2} .$$

Now, combining both these fractions, we easily obtain that

$$\rho_1 + (1 - \rho_1)\frac{\rho_2 + \widehat{\rho}_2}{1 + \rho_2 + \widehat{\rho}_2} = \frac{\rho + \widehat{\rho}_2}{\rho_2 + \widehat{\rho}_2 + 1} = \mathsf{P}_B .$$

Note that in the analysis above we implicitly used the PASTA property, which enables to equate the fraction of class-2 arrivals which observe a busy server occupied by another class-2 customer and the fraction of time when the server is occupied by class-2 customers.

Problem 7.2 Verify formula (7.5.13).

Problem 7.3 Check that conditions (7.5.11) and (7.5.12) imply $\mathsf{P}_B \le \rho$.

7.6 Explicit Expressions for the Busy Probability

Following [3], we now consider a few m-server *single-class retrial systems* (with retrial rate μ_0) for which the probability $\widehat{\mathsf{P}}_B$ can be found in an explicit form.

For the bufferless retrial $M/G/m/0$-type system (which is useful for studying optical networks [5, 6]) with input rate λ and service rate μ, one can use the celebrated *Erlang loss formula* to obtain

$$\widehat{\mathsf{P}}_B = \frac{\rho_0^m}{m!}\left[\sum_{n=0}^{m}\frac{\rho_0^n}{n!}\right]^{-1},$$

where $\rho_0 := (\lambda + \mu_0)/\mu$.

Exponential service time. Assume that the primary system of the m-server system $\widehat{\Sigma}$ is a $M/M/m/N$ system, with buffer size N and aggregated input rate $\lambda + \mu_0$. Note that, since the ordering (7.5.5) is reflexive in the case of the exponential distribution, we can set $\widehat{\mu} = \mu$ [4]. It is well-known (see for instance, [7]) that in this case the stationary probability that the primary system of $\widehat{\Sigma}$ is full is given by

$$\widehat{\mathsf{P}}_B = \frac{\rho_0^{m+N}}{m!\,m^N}\mathsf{P}_0\,,$$

and the stability condition (7.5.4) takes the form

$$\frac{\rho_0^{m+N}}{m!\,m^N}\mathsf{P}_0 < \frac{\mu_0}{\lambda + \mu_0}\,, \tag{7.6.1}$$

where

$$\mathsf{P}_0 = \left[\sum_{n=0}^{m}\frac{\rho_0^n}{n!} + \frac{\rho_0^m}{m!}\sum_{n=1}^{N}\left[\frac{\rho_0}{m}\right]^n\right]^{-1}.$$

In particular, in a single-server $M/M/1/N$ system, the above condition reduces to

$$\frac{\rho_0^{N+1}}{\sum_{n=0}^{N+1}\rho_0^n} < \frac{\mu_0}{\lambda + \mu_0}\,,$$

which is equivalent to the stability condition established in [8].

In a system with *recovery probability* p, the fraction $\lambda(1-p)$ of new arrivals goes directly to the orbit. Then, in the $M/M/m/0$ system with recovery probability $p = 1$, condition (7.6.1) reduces to

$$\frac{\rho_0^m}{m!}\left[\sum_{n=0}^{m}\frac{\rho_0^n}{n!}\right]^{-1} < \frac{\mu_0}{\lambda + \mu_0}\,,$$

which is equivalent to the stability condition derived in [9]. If the recovery probability $p < 1$, then the stability condition in [9] can be written as

$$(1-p)\lambda + (p\lambda + \mu_0)\widehat{\mathsf{P}}_B < \mu_0\,, \tag{7.6.2}$$

where

$$\widehat{\mathsf{P}}_B = \frac{[p\lambda + \mu_0]^m}{\mu^m \, m!} \left[\sum_{n=0}^{m} \frac{[p\lambda + \mu_0]^n}{\mu^n \, n!} \right]^{-1} .$$

Apparently, $\widehat{\mathsf{P}}_B$ represents the stationary *loss probability* in the primary system of $\widehat{\Sigma}$, with a Poisson arrival process with rate $p\lambda + \mu_0$. Then the first term $(1 - p)\lambda$ on the left-hand side of inequality (7.6.2) is the arrival rate of primary customers going directly to the orbit, while the second term is the rate of rejected retrial customers.

Hyperexponential service time. For the *hyperexponential service time* distribution function

$$F(x) = 1 - \sum_{i=1}^{n} p_i e^{-\mu_i x}, \quad x \geq 0 ,$$

(with $\sum_{i=1}^{n} p_i = 1$), the failure rate is given by

$$r(x) = \frac{\sum_{i=1}^{n} p_i \mu_i e^{-\mu_i x}}{\sum_{i=1}^{n} p_i e^{-\mu_i x}} ,$$

and is a decreasing function, satisfying

$$\lim_{x \to \infty} r(x) = \min(\mu_1, ..., \mu_n) .$$

Hence, according to (7.5.6), we can set

$$\widehat{\mu} = \min(\mu_1, ..., \mu_n) .$$

7.7 Notes

Retrial queues are broadly used for modeling many practical problems, such as those related to call centers [10], computer networks, cellular networks, medium access protocols in wired and wireless networks, and optical switching networks, but also induce a significant independent theoretical interest. A more extensive review of retrial queueing literature can be found, for example, in [11–16]. Specific features of the retrial queue model, such as the arrival and service processes, the retrial discipline, customer patience, the number of servers and the number of customer classes, etc., may dramatically complicate their analysis.

The detailed proof of Theorem 7.4 uses monotonicity results from [4], which in turn are based on the results presented in [17, 18]. If an explicit expression for $\widehat{\mathsf{P}}_B$ is not available, one can rely on an estimate of $\widehat{\mathsf{P}}_B$ which could be obtained by simulation quickly and with a high accuracy in a lot of cases. Moreover, as one is primarily interested in the values of $\widehat{\mathsf{P}}_B$ for the boundary of the stability region, the value of $\widehat{\mathsf{P}}_B$ near the stability boundary is quite high, and its estimation is relatively easy (for more details, see [19, 20]).

The stability condition obtained in Theorem 7.5 also can be applied for *Gamma service time distribution* with shape parameter $s \in (0, 1)$ [3], and also for a broad class of service time distributions with *increasing failure rates* [3, 4, 21, 22]. (We note that the operator min in the rhs of (7.5.6) is mistakenly omitted in paper [2].) At first glance, the assumptions of Theorem 7.5 seem to be rather restrictive to be useful. However, as we show above, conditions (7.5.4) do yield the stability criterion in case of symmetrical orbits, and they are also valid for a wider class of queueing systems. (See for instance [23], where the stability of the $GI/M/1/0$ retrial system has been analyzed.)

An interesting example of the existence of a local stability of a two-dimensional Markov chain is analyzed in [24].

We would also like to point out that, in general, there is a difference between a *non-idling* service discipline, where the server(s) can not be idle while there are customers waiting for service [25, 26], and a *work-conserving* discipline, where work can not be created/destroyed *within the system* [27, 28]. The service discipline in the N-model discussed in Sect. 6.2 and in the retrial systems that we study in Chaps. 7 and 8 are work-conserving, but not non-idling. However, we could also interpret the idle time of the server(s), while the system is non-empty, as work created within the system, making the service discipline non-work-conserving.

References

1. Morozov, E., Rumyantsev, A., Dey, S., Deepak, T.G.: Performance analysis and stability of multiclass orbit queue with constant retrial rates and balking. Perf. Eval. 134 (2019). https://doi.org/10.1016/j.peva.2019.102005
2. Avrachenkov, K., Morozov, E., Steyaert, B.: Sufficient stability conditions for multiclass constant retrial rate systems. Queueing Syst. **82**(1–2), 149–171 (2016)
3. Avrachenkov, K., Morozov, E.: Stability analysis of GI/GI/c/K retrial queue with constant retrial rate. Math. Meth. Oper. Res. **79**, 273–291 (2014)
4. Whitt, W.: Comparing counting processes and queues. Adv. Appl. Prob. **13**, 207–220 (1981)
5. Wong, E.W.M., Andrew, L.L.H., Cui, T., Moran, B., Zalesky, A., Tucker, R.S., Zukerman, M.: Towards a bufferless optical internet. J. Lightw. Tech. **27**, 2817–2833 (2009)
6. Yao, S., Xue, F., Mukherjee, B., Yoo, S.J.B., Dixit, S.: Electrical ingress buffering and traffic aggregation for optical packet switching and their effect on TCP-level performance in optical mesh networks. IEEE Commun. Mag. **40**(9), 66–72 (2002)
7. Kleinrock, L.: Queueing Systems. Volume 1: Theory. Wiley, New York (1975)
8. Avrachenkov, K., Yechiali, U.: Retrial networks with finite buffers and their application to Internet data traffic. Prob. Engin. Inf. Sci. **22**, 519–536 (2008)
9. Artalejo, J.R., Gómez-Corral, A., Neuts, M.F.: Analysis of multiserver queues with constant retrial rate. Eur. J. Oper. Res. **135**, 569–581 (2001)
10. Aguir, S., Karaesmen, F., Aksin, O.Z., Chauvet, F.: The impact of retrials on call center performance. OR Spectrum **26**, 353–376 (2004)
11. Artalejo, J.R.: Accessible bibliography on retrial queues. Math. Comput. Mod. **30**(3–4), 1–6 (1999)

12. Artalejo, J.R.: A classified bibliography of research on retrial queues: progress in 1990–1999. Top **7**(2), 187–211 (1999)
13. Avrachenkov, K., Nain, P., Yechiali, U.: A retrial system with two input streams and two orbit queues. Queueing Syst. **77**(1), 1–31 (2014)
14. Falin, G.I.: A survey of retrial queues. Queueing Syst. **7**(2), 127–167 (1990)
15. Falin, G.I., Templeton, J.G.C.: Retrial Queues. Chapman and Hall, London (1997)
16. Yang, T., Templeton, J.G.C.: A survey on retrial queues. Queueing Syst. **2**(3), 201–234 (1987)
17. Sonderman, D.: Comparing multi-server queues with finite waitng rooms, I: Same number of servers. Adv. Appl. Prob. **11**(2), 439–447 (1979)
18. Sonderman, D.: Comparing multi-server queues with finite waiting rooms, II: Different number of servers. Adv. Appl. Prob. **11**(2), 448–455 (1979)
19. Avrachenkov, K., Goricheva, R., Morozov, E.: Verification of stability region of a retrial queuing system by regenerative method. In: International Proceedings Modern Probabilistic Methods for Analysis and Optimization of Information and Telecommunication Networks, pp. 22–28. Minsk (2011)
20. Avrachenkov, K., Morozov, E., Nekrasova, R., Steyaert, B.: Stability analysis and simulation of N-class retrial system with constant retrial rates and Poisson inputs. Asia-Pacific J. Oper. Res. 31(2) (2014)
21. Barlow, R.E., Proschan, F.: Mathematical Theory of Reliability. Classics in Applied Mathematics (SIAM). Wiley, New York (1987)
22. Müller A., Stoyan D.: Comparisons Methods for Stochastic Models and Risks. Wiley, Hoboken (2002)
23. Lillo, R.E.: A G/M/1-queue with exponential retrial. TOP **4**, 99–120 (1996)
24. Adan, I., Foss, S., Weiss, G.: Local stability in a transient Markov chain. Stat. Prob. Lett. **165**, 1–6 (2020)
25. Dai, J.: On positive Harris recurrence of multiclass queueing networks: a unified approach via fluid limit models. Ann. Appl. Prob. **5**, 49–77 (1995)
26. Dai, J., Hasenbein, J., Kim, B.: Stability of join-the-shortest-queue networks. Queueing Syst. **57**, 129–145 (2007)
27. Boxma, O.,J., Groenendijk, W.P.: Peseudo-conservation laws in cyclic-service systems. J. Appl. Prob. **24**(4), 949–964 (1987)
28. Wolff, R.W.: Work-conserving priorities. J. Appl. Prob. **7**(2), 327–337 (1970)

Chapter 8
Systems with State-Dependent Retrial Rates

In this chapter, we extend the analysis developed in Chap. 7 to more complicated retrial systems, in which the retrial rate of each orbit depends on a *binary state* of the other orbits. This setting is well-motivated and suited for modelling wireless multiple-access systems, see Sect. 8.4.

It is worth mentioning that, as in Chap. 7, our proof is based on a coupling argument, relating the process of retrial attempts from an orbit to an independent Poisson process in which the rate corresponds to the current state of the other orbits. In turn, this then allows to apply the PASTA property to deduce both the stability condition as well as some explicit expressions or bounds for stationary probabilities that characterize the retrial system.

8.1 Model Description

As in [1, 2], we study the following modification of the single-server retrial queueing model with K classes *without balking*, that was considered in Chap. 7: maintaining all other assumptions and notation, we now assume that the attempts from (a non-empty) orbit i follow an exponential distribution with a rate depending on i and the current status of other orbits: *idle or non-idle*. For each orbit i, we define the set $\mathcal{G}(i) = \{J_n^{(i)}\}$ of K-dimensional vectors

$$J_n^{(i)} = \{j_{n,1}^{(i)}, \ldots, j_{n,i-1}^{(i)}, 1, j_{n,i+1}^{(i)}, \ldots, j_{n,K}^{(i)}\},$$

with binary components $j_{n,k}^{(i)} \in \{0, 1\}$, and where the ith component always satisfies $j_{n,i}^{(i)} = 1$. We assume that $\mathcal{G}(i)$ is an ordered set (say, in a lexicographical order), and index n denotes the nth element of the set. Note that in general, the cardinality of this set is given by $|\mathcal{G}(i)| = 2^{K-1}$. Each vector $J_n^{(i)}$ is called the *configuration*

E. Morozov and B. Steyaert, *Stability Analysis of Regenerative Queueing Models*,
https://doi.org/10.1007/978-3-030-82438-9_8

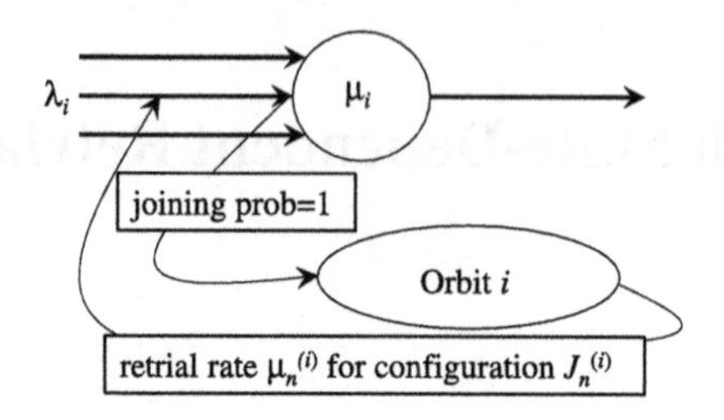

Fig. 8.1 A K-class single-server retrial system: a class-i customer observing a busy server joins orbit i and retries with rate $\mu_n^{(i)}$ depending on the current configuration $J_n^{(i)}$

and has the following interpretation: if the class-k orbit is non-idle, then we set $j_{n,k}^{(i)} = 1$, otherwise, $j_{n,k}^{(i)} = 0$. For a given configuration $J_n^{(i)}$, we denote by $\mu_n^{(i)}$ the retransmission rate of orbit i, a setting that is depicted in Fig. 8.1. Hence, the set $\mathcal{G}(i)$ contains all possible different configurations of the other orbits 'observed from busy orbit i'. Also we define the set of rates

$$\mathcal{M}_i = \{\mu_n^{(i)} : J_n^{(i)} \in \mathcal{G}(i)\} ,$$

of all configurations belonging to $\mathcal{G}(i)$, $1 \le i \le K$. In general, orbit i has different retransmission rates for different configurations, but it is of course possible that the set $\mathcal{M}_i$ contains repetitive elements, say $\mu_k^{(i)} = \mu_l^{(i)}$. This means that the retrial rate of orbit i is *insensitive* to switching between configurations $J_k^{(i)}$ and $J_l^{(i)}$. In order to describe the *regenerative structure* of this retrial system, we first define the basic stochastic processes. Let $N_i(t)$ be the number of class-i customers blocked in orbit i, and $W_i(t)$ be the remaining work in orbit i, at instant t. Define

$$N(t) = \sum_{i=1}^{K} N_i(t), \quad t \ge 0 ,$$

and let $Q(t) = 1$ if the server is busy at instant t, and $Q(t) = 0$, otherwise. Finally, let

$$X(t) = N(t) + Q(t), \; t \ge 0 ,$$

and denote

$$X(t_k^-) = N(t_k^-) + Q(t_k^-) =: N_k + Q_k =: X_k, \quad k \ge 1 .$$

We will consider a zero initial state, $X_1 = 0$, in which case the regeneration instants of $\{X(t)\}$ are given by

$$T_0 = 0, \; T_{n+1} = \inf_{k \ge 1}\left(t_k > T_n : X_k = 0\right) , \; n \ge 0 .$$

8.2 Necessary Stability Condition

Let us introduce the *maximal* possible retrial rate from orbit i as

$$\widehat{\mu}_i = \max_{n: J_n^{(i)} \in \mathcal{G}(i)} \mu_n^{(i)}, \ 1 \le i \le K .$$

The proof of the following necessary stability condition is in part similar to the proof of Theorem 7.1.

Theorem 8.1 *If the system under consideration is positive recurrent, then*

$$\rho < \min_{1 \le i \le K} \left[\frac{\widehat{\mu}_i}{\lambda_i + \widehat{\mu}_i} \right] . \tag{8.2.1}$$

Proof Adopting the same notation as before (in particular, indicators $\mathsf{1}_n^{(i)}$, see (7.4.2)), the time $V_i(t)$ that is required to process all class-i customers arriving in the interval $[0, \ t]$ is the sum

$$V_i(t) = \sum_{n=1}^{A(t)} \mathsf{1}_n^{(i)} S_n^{(i)}, \ 1 \le i \le K , \tag{8.2.2}$$

which contains iid summands. Recall that $\mathsf{E}\mathsf{1}^{(i)} = p_i$ and $\lambda_i = \lambda p_i$. As in Chap. 7, making use of the SLLN and balance equation (7.2.6), we obtain the stationary probability that the *server is occupied by a class-i customer* as the following limit (analogous to (7.2.7)):

$$\begin{aligned} \mathsf{P}_b^{(i)} &= \lim_{t \to \infty} \frac{B_i(t)}{t} = \lim_{t \to \infty} \frac{V_i(t)}{t} \\ &= \lim_{t \to \infty} \frac{\sum_{n=1}^{A(t)} \mathsf{1}_n^{(i)} S_n^{(i)}}{A(t)} \frac{A(t)}{t} = \frac{\lambda_i}{\mu_i} = \rho_i , \quad 1 \le i \le K . \end{aligned} \tag{8.2.3}$$

Then the stationary *busy probability of the server* is given by

$$\mathsf{P}_B = \sum_{i=1}^{K} \mathsf{P}_b^{(i)} = \sum_{i=1}^{K} \rho_i =: \rho . \tag{8.2.4}$$

As before, due to the positive recurrence of the involved queueing processes, $Q_n \Rightarrow Q$, and in view of the PASTA property, the stationary busy probability equals (see also (7.2.13))

$$\mathsf{P}(Q = 1) = \mathsf{E}Q = \mathsf{P}_B = \rho . \tag{8.2.5}$$

Denote

$$A_i^{(0)}(t) = \sum_{n=1}^{A(t)} 1_n^{(i)} Q_n ,$$

the number of blocked class-i customers in the interval $[0, t]$. Now, due to the positive recurrence, the mean cycle increment of the process $\{A_i^{(0)}(t)\}$ with regenerative increments, is upper bounded by $\mathsf{E}\theta < \infty$, and we find from (2.1.7) that, for $1 \le i \le K$,

$$\begin{aligned}\lim_{t\to\infty} \frac{1}{t} A_i^{(0)}(t) &= \lim_{t\to\infty} \frac{A(t)}{t} \lim_{t\to\infty} \frac{1}{A(t)} \sum_{n=1}^{A(t)} 1_n^{(i)} Q_n \\ &= \lambda \mathsf{E}[1^{(i)} Q] = \lambda p_i \rho = \lambda_i \rho ,\end{aligned} \tag{8.2.6}$$

where we have applied (8.2.5) and the independence between $1^{(i)}$ and the (stationary) state Q of the server. Define $\widehat{\mathbb{D}}_n^{(i)} = \{\widehat{D}_n^{(i)}(t),\ t \ge 0\}$ as the Poisson process with rate $\mu_n^{(i)}$, implying (see (7.2.14), (7.2.17))

$$\lim_{t\to\infty} \frac{1}{t} \widehat{D}_n^{(i)}(t) = \mu_n^{(i)} . \tag{8.2.7}$$

We may consider the process $\widehat{\mathbb{D}}_n^{(i)}$ as being generated by the attempts from a (non-idle) orbit i, if the system would permanently remain in configuration $J_n^{(i)}$. For each i, introduce the family of *independent* Poisson processes

$$\mathcal{D}_i = \{\widehat{\mathbb{D}}_n^{(i)} : J_n^{(i)} \in \mathcal{G}(i)\} ,$$

corresponding to all possible configurations $J_n^{(i)}$ forming the set $\mathcal{G}(i)$. The subsequent analysis is similar to the one that has been developed in Theorem 7.1, where by means of resampling and coupling, the process $\{D_i(t)\}$ of the real retrial attempts from orbit i, was represented as a *subsequence* of events of the corresponding Poisson process $\widehat{\mathbb{D}}_i$. The key novel element of the current analysis is that we now connect the process $\{D_i(t)\}$ with the Poisson process $\widehat{\mathbb{D}}_n^{(i)}$, *during time periods when the system resides in configuration* $J_n^{(i)}$. Let $\{u_n^{(i)}(k),\ k \ge 1\}$ be the renewal instants of the process $\widehat{\mathbb{D}}_n^{(i)}$.

This construction is shown in Fig. 8.2, where (in order to simplify the notation) we have omitted index i, and where the quantity Δ, at the *synchronization points* $\{v_n\}$, represents the (random) gap size in the corresponding remaining retrial time process, caused by the resampling procedure. In particular, when the system is in configuration $J_n^{(i)}$, then we resample the process $\widehat{\mathbb{D}}_n^{(i)}$ while the renewal processes describing other configurations continue to evolve as before, see the process $\widehat{\mathbb{D}}_k^{(i)}$ describing configuration $J_k^{(i)}$ in Fig. 8.2. (For sake of illustration, in Fig. 8.2 we describe the dynamics of retrials by the corresponding remaining renewal times $\xi(t),\ \widehat{\xi}_k(t),\ \widehat{\xi}_n(t)$.)

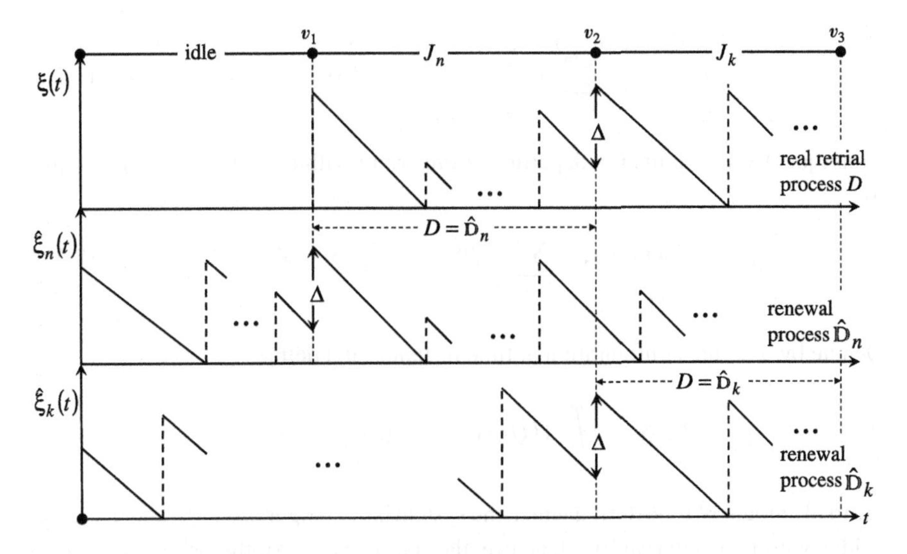

Fig. 8.2 The remaining retrial time $\xi(t)$ in the original process of attempts and the remaining renewal times $\widehat{\xi}_k(t)$, $\widehat{\xi}_n(t)$ in the associated renewal processes $\widehat{\mathbb{D}}_n$ and $\widehat{\mathbb{D}}_k$, respectively

Analogously, this procedure is applied to all other orbits. Therefore, the instants of the actual retrials constitute a *subsequence* of the renewal points of the corresponding processes *with resampling* from the family $\mathcal{D}_i$. This resampling procedure does not change the distribution of the Poisson processes from the family $\mathcal{D}_i$, and we will maintain the notation $\widehat{\mathbb{D}}_n^{(i)}$ for the resampled process, and the notation $\{u_n^{(i)}(k),\ k \geq 1\}$ for its instants. Now, for each i, denote by

$$Q(u_n^{(i)}(k)^-) = \mathcal{Q}_n^{(i)}(k),\ \ J_i(u_n^{(i)}(k)^-) = J_n^{(i)}(k),\ \ n \geq 1\ .$$

In particular, $J_n^{(i)}(k)$ is the (random) configuration of the system at instant $u_n^{(i)}(k)$. Note that the i-orbit is not idle in any configuration $J_n^{(i)}$, and therefore the equality

$$1\big(\mathcal{Q}_n^{(i)}(k) = 0,\ \ J_n^{(i)}(k) = J_n^{(i)}\big) = 1\ ,$$

indicates that the instant of the kth event of the process $\widehat{\mathbb{D}}_n^{(i)}$ corresponds to a *successful attempt* of a class-i customer while the system is in configuration $J_n^{(i)}$. On the other hand,

$$1\big(\mathcal{Q}_n^{(i)}(k) = 1,\ \ J_n^{(i)}(k) = J_n^{(i)}\big) = 1\ ,$$

implies that the server is busy, and the attempt to access the server is *unsuccessful* and can indeed be ignored. Then the number of successful attempts $D_n^{(i)}(t)$ from orbit i within the time interval $[0,\ t]$, when the system has configuration $J_n^{(i)}$, is defined as

$$D_n^{(i)}(t) = \sum_{k=1}^{\widehat{D}_n^{(i)}(t)} 1\big(\mathcal{Q}_n^{(i)}(k) = 0,\ J_n^{(i)}(k) = J_n^{(i)}\big) . \tag{8.2.8}$$

Consequently, the number of departures from orbit i within this time interval is equal to

$$D_i(t) = \sum_{n: J_n^{(i)} \in \mathcal{G}(i)} D_n^{(i)}(t),\ \ 1 \le i \le K . \tag{8.2.9}$$

Denote by $J(t)$ the configuration at time instant t, and define

$$\mathbb{T}_n^{(i)}(t) = \int_0^t 1\big(Q(u) = 0,\ J(u) = J_n^{(i)}\big)du ,$$

the total amount of time during which the system has configuration $J_n^{(i)}$ and the server is idle, within the interval $[0,\ t]$. Hence, the stationary probability $\mathsf{P}_n^{(i)}$ that *server is idle and the system configuration is* $J_n^{(i)}$ can be obtained as the *time-average* limit

$$\mathsf{P}_n^{(i)} = \lim_{t\to\infty} \frac{\mathbb{T}_n^{(i)}(t)}{t} . \tag{8.2.10}$$

On the other hand, due to the construction of the process $\widehat{\mathbb{D}}_n^{(i)}$, we can invoke PASTA to conclude that the following *event-average* limit coincides with the limit in (8.2.10):

$$\lim_{t\to\infty} \frac{1}{\widehat{D}_n^{(i)}(t)} \sum_{k=1}^{\widehat{D}_n^{(i)}(t)} 1\big(\mathcal{Q}_n^{(i)}(k) = 0,\ J_n^{(i)}(k) = J_n^{(i)}\big) = \mathsf{P}_n^{(i)} . \tag{8.2.11}$$

Therefore, from (8.2.7), (8.2.8) and (8.2.11), we find that, w.p.1 (see also (7.2.18)),

$$\lim_{t\to\infty} \frac{D_n^{(i)}(t)}{t} = \mathsf{P}_n^{(i)} \mu_n^{(i)} . \tag{8.2.12}$$

Now, using (8.2.9) and (8.2.12), we obtain

$$\lim_{t\to\infty} \frac{D_i(t)}{t} = \lim_{t\to\infty} \frac{1}{t} \sum_{n: J_n^{(i)} \in \mathcal{G}(i)} D_n^{(i)}(t) = \sum_{n: J_n^{(i)} \in \mathcal{G}(i)} \mu_n^{(i)} \mathsf{P}_n^{(i)} . \tag{8.2.13}$$

Because

$$N_i(0^-) + A_i^{(0)}(t) = N_i(t) + D_i(t) ,\ \ 1 \le i \le K ,$$

and $N_i(t) = o(t)$, we then obtain from (8.2.4), (8.2.6) and (8.2.13) the following relation for each orbit i:

$$\lambda_i \mathsf{P}_B = \lambda_i \rho = \sum_{n:J_n^{(i)} \in \mathcal{G}(i)} \mu_n^{(i)} \mathsf{P}_n^{(i)} . \tag{8.2.14}$$

As indicated before, in each configuration $J_n^{(i)}$, orbit i is busy, and an important observation then is that the expression

$$\sum_{n:J_n^{(i)} \in \mathcal{G}(i)} \mathsf{P}_n^{(i)} = \mathsf{P}_{0,b}^{(i)} , \tag{8.2.15}$$

is equal to the stationary probability that *the server is idle and the orbit i is busy.* Note that this probability, for each orbit i, is related to the stationary *idle server* probability P_0 as follows:

$$\mathsf{P}_0 = \mathsf{P}_{0,b}^{(i)} + \mathsf{P}_{0,0}^{(i)} , \tag{8.2.16}$$

where $\mathsf{P}_{0,0}^{(i)}$ is the stationary probability that *both the server and orbit i are idle*, also see (7.2.3) and (7.2.4) (with $C = b_k \equiv 1$). Further denote by π_0 the stationary probability that the *system is completely empty*. Due to the positive recurrence of the system, we find that

$$\pi_0 = \lim_{t \to \infty} \frac{1}{t} \int_0^t 1(X(u) = 0)du = \frac{\mathsf{E}I_0}{\mathsf{E}T} = \frac{1}{\mathsf{E}\theta} > 0 , \tag{8.2.17}$$

where I_0 denotes the idle period of the server preceding a regeneration instant. To deduce (8.2.17), we have applied Wald's identity $\mathsf{E}T = \mathsf{E}\tau\, \mathsf{E}\theta$, and also relied on the memoryless property of the exponential interarrival time, implying $\mathsf{E}I_0 = \mathsf{E}\tau$. Because of

$$\mathsf{P}_{0,0}^{(i)} \geq \pi_0 > 0 ,$$

then it follows from (8.2.16) that

$$\mathsf{P}_{0,b}^{(i)} < \mathsf{P}_0 = 1 - \mathsf{P}_B .$$

Then, relying on (8.2.4) and (8.2.14)–(8.2.16), the following inequalities hold:

$$\lambda_i \rho \leq \widehat{\mu}_i \mathsf{P}_{0,b}^{(i)} < \widehat{\mu}_i \mathsf{P}_0 = \widehat{\mu}_i (1 - \rho), \quad 1 \leq i \leq K ,$$

indeed leading to (8.2.1). □

Remark 8.1 Relation (8.2.17) is a particular case of the PASTA property: we equate the fraction of idle time, $\mathsf{E}I_0/\mathsf{E}T$, and the fraction of customers, $1/\mathsf{E}\theta$, observing an idle system upon arrival.

Define the *minimal* possible retrial rate from orbit i,

$$\mu_i^0 = \min_{n:J_n^{(i)} \in \mathcal{G}(i)} \mu_n^{(i)}, \ 1 \le i \le K .$$

The proved conditions imply some useful bounds for the related stationary probabilities. For instance, from (8.2.14)–(8.2.16), we obtain the following upper bound:

$$\mathsf{P}_{0,b}^{(i)} = \sum_{n:J_n^{(i)} \in \mathcal{G}(i)} \mathsf{P}_n^{(i)} \le \sum_{n:J_n^{(i)} \in \mathcal{G}(i)} \mathsf{P}_n^{(i)} \frac{\mu_n^{(i)}}{\mu_i^0} = \frac{\lambda_i \rho}{\mu_i^0}, \quad (8.2.18)$$

and, also in view of (8.2.4) and (8.2.14), a lower bound for the probability $\mathsf{P}_{0,0}^{(i)}$:

$$\mathsf{P}_{0,0}^{(i)} = 1 - \mathsf{P}_B - \mathsf{P}_{0,b}^{(i)} \ge 1 - \rho\big(1 + \frac{\lambda_i}{\mu_i^0}\big), \quad 1 \le i \le K .$$

If the retrial rate is insensitive to the configurations and depends solely on the orbit number, i.e.,

$$\mu_n^{(i)} \equiv \mu_i \ \text{ for all } \ J_n^{(i)} \in \mathcal{G}(i) ,$$

then the system becomes a conventional *constant retrial rates multiserver system*, see Chap. 7 and [3, 4]. Moreover, in this case inequalities (8.2.18) transform to the following explicit expressions:

$$\mathsf{P}_{0,b}^{(i)} = \frac{\lambda_i}{\mu_i}\rho, \ 1 \le i \le K ,$$

which coincide with (7.2.3) if $b_k \equiv 1$; also see [1, 5].

8.3 Sufficient Stability Condition

The complete proof of the sufficient stability condition for the current system is indeed quite similar to the one that was presented for the multiclass retrial system in Theorem 7.3. For this reason, in this section we confine ourselves to outlining the corresponding proof.

We first refer to construction (7.4.2) of the service times. Next, we construct a *dominating buffered system*, denoted by $\widehat{\Sigma}$, as follows. Using a coupling technique, we use the same interarrival times in both systems (with generic time τ). Then, in system $\widehat{\Sigma}$, for each class-i customer, we assign, in addition to the original service time $S^{(i)}$, an extra exponential service time

$$\zeta_i =_{st} \min(\tau, \xi_i^o) ,$$

where ξ_i^o is an exponential random variable with *minimal* parameter μ_i^0. In other words, ξ_i^o is the distance between the *'slowest'* attempts from orbit i. The variable ζ_i has rate $\lambda + \mu_i^0$, and is an upper bound for the idle time of the server after a departure, provided that orbit i is busy. Then it follows that the generic service time $\widehat{S}^{(i)}$ in the system $\widehat{\Sigma}$ satisfies the stochastic equality

$$\widehat{S}^{(i)} =_{st} S^{(i)} + \zeta_i, \; i = 1, \ldots, K .$$

Furthermore, invoking the corresponding coupling argument and by induction, we can show that (i) it is possible to serve customers in both systems in the *same order*, and that (ii) a customer leaves the system $\widehat{\Sigma}$ *not earlier* than the same customer in the original system [6]. This monotonicity property implies that the remaining work in the original system is dominated by the remaining work in system $\widehat{\Sigma}$. This observation in turn means that the positive recurrence of system $\widehat{\Sigma}$ implies positive recurrence of the original system. On the other hand, a standard *negative drift* condition for the buffered system $\widehat{\Sigma}$ to be positive recurrent is $\mathsf{E}\widehat{S} < \mathsf{E}\tau$. In view of $\mathsf{E}S^{(i)} = 1/\mu_i, \; p_i = \lambda_i/\lambda$, then representation (7.4.2) allows us to write condition $\mathsf{E}\widehat{S} < \mathsf{E}\tau$ as follows:

$$\begin{aligned} \mathsf{E}\widehat{S} &= \sum_{i=1}^{K} p_i [\mathsf{E}S^{(i)} + \mathsf{E}\zeta_i] = \sum_{i=1}^{K} \frac{\lambda_i}{\lambda} \Big[\frac{1}{\mu_i} + \frac{1}{\lambda + \mu_i^0} \Big] \\ &= \frac{1}{\lambda} \sum_{i=1}^{K} \Big[\rho_i + \frac{\lambda_i}{\lambda + \mu_i^0} \Big] < \frac{1}{\lambda} . \end{aligned}$$

The above analysis therefore leads to the following statement:

Theorem 8.2 *The zero initial state retrial system under consideration is positive recurrent if the following condition holds:*

$$\rho + \sum_{1 \le i \le K} \frac{\lambda_i}{\mu_i^0 + \lambda} < 1 . \qquad (8.3.1)$$

8.4 Notes

The retrial systems with *state-dependent retrial rates*, or with *coupled orbits* are studied, in particular, in [7–9]. In such a system, users transmit packets to a common destination node, and the orbit queues play the role of relay nodes to retransmit blocked packets, see for instance [10]. There is indeed a need for developing cognitive radio communication systems, to solve a spectrum of under-utilization problems [11]. In modern cognitive radio communication systems, a wireless node is capable to access the status of the operational environment, and this opens a range of possibilities to dynamically adjusts the operational parameters (such as the retransmission rates) to achieve close-to-full spectrum utilization, also see [12]. Another application that we would like to mention is that of cellular networks, where the available retransmission rate (in a particular cell) decreases as the number of users in the neighboring cells increase [13]. A similar effect also arises in processor sharing models [14].

A numerical analysis of the 'gap' between the necessary stability condition (8.2.1) and sufficient condition (8.3.1) is presented in [1, 4]. A multiclass system with state-dependent retrial rates with a *non-reliable server* and with a *preemptive-repeat* and *preemptive-resume* service policy has been considered in [15, 16].

References

1. Morozov, E., Morozova, T.: Analysis of a generalized retrial system with coupled orbits. In: Proceeding 23rd Conference of Open Innovations Association (FRUCT), pp. 253–260. Bologna (2018)
2. Morozov, E., Morozova, T.: A coupling-based analysis of a multiclass retrial system with state-dependent retrial rates. In: Proceedings 14th International Conference on Queueing Theory and Network Applications (QTNA 2019, LNCS 11688), pp. 34–50. Ghent (2019)
3. Avrachenkov, K., Morozov, E., Steyaert, B.: Sufficient stability conditions for multiclass constant retrial rate systems. Queueing Syst. **82**(1–2), 149–171 (2016)
4. Morozov, E., Phung-Duc, T.: Regenerative analysis of two-way communication orbit-queue with general service time. In: Proceedings International Conference Queueing Theory and Network Applications (QTNA 2018, LNCS 10932), pp. 22–32. Tsukuba (2018)
5. Morozov, E., Dimitriou, I.: Stability analysis of a multiclass retrial system with coupled orbit queues. In: Proceedings 14th European Workshop, Computer Performance Engineering (EPEW 2017), pp. 85–98. Berlin (2017)
6. Morozov, E., Rumyantsev, A., Dey, S., Deepak, T.G.: Performance analysis and stability of multiclass orbit queue with constant retrial rates and balking. Perf. Eval. **134**, (2019). https://doi.org/10.1016/j.peva.2019.102005
7. Dimitriou, I.: A two class retrial system with coupled orbit queues. Prob. Eng. Infor. Sci. **31**(2), 139–179 (2017)
8. Dimitriou, I.: A queueing system for modeling cooperative wireless networks with coupled relay nodes and synchronized packet arrivals. Perf. Eval. **114**, 16–31 (2017)
9. Dimitriou, I.: Modeling and analysis of a relay-assisted cooperative cognitive network. In: Proceedings Analytical and Stochastic Modelling Techniques and Applications (LNCS), pp. 47–62. Newcastle (2017)
10. Sadek, A., Liu, K., Ephremides, A.: Cognitive multiple access via cooperation: protocol design and performance analysis. IEEE Trans. Infor. Theor. **53**(10), 3677–3696 (2007)
11. Mitola, J., Maguire, G.Q.: Cognitive radio: making software radios more personal. IEEE Pers. Commun. **6**(4), 13–18 (1999)
12. Borst, S., Jonckheere, M., Leskelä, L.: Stability of parallel queueing systems with coupled service rates. Discr. Ev. Dyn. Syst. **18**(4), 447–472 (2008)
13. Bonald, T., Borst, S., Hegde, N., Proutière, A.: Wireless data performance in multi-cell scenarios. In: Proceedings ACM Sigmetrics/Performance '04, pp. 378–388. New York (2004)
14. Bonald, T., Massoulié, L., Proutière, A., Virtamo, J.: A queueing analysis of max-min fairness, proportional fairness and balanced fairness. Queueing Syst. **53**(1–2), 65–84 (2006)
15. Dimitriou, I., Morozov, E., Morozova, T.: A multiclass retrial system with coupled orbits and service interruptions: Verification of stability conditions. In: Proceedings of the 24th Conference of Open Innovations Association, FRUCT, 2019 ISSN: 2305-7254 eISSN: 2343-0737, vol. 24, pp. 75–81 (2019)
16. Morozov, E., Morozova, T.: The remaining busy time in a retrial system with unreliable servers. Proceedings International Conference on Distributed Computer and Communication Networks (DCCN 2020, LNCS 12563), pp. 555–566. Moscow (2020)

Chapter 9
A Multiclass Multiserver System with Classical Retrials

The purpose of this chapter is to derive the stability condition for another modification of the K-class retrial system considered in Chap. 7, with total input rate $\lambda = \sum_{i=1}^{K} \lambda_i$ (with no balking), with a *finite buffer* of size $G \geq 0$, and with m identical servers. Adopting most of the notations used in Chap. 7, we now consider a *classical* retrial discipline, in which case a class-i customer, observing a completely occupied primary system (servers and buffer) upon arrival, joins the (virtual) i-orbit and then generates *independent* retrial attempts with rate $\mu_0^{(i)}$ to access the primary system. The proof of the stability condition is borrowed from paper [1]. As always, regenerations of this system occur when a new arrival observes a completely empty system, see (7.1.2).

The stability analysis exploits the property that the service discipline in this system is *asymptotically non-idling* [2]. (See also a comment in Sect. 10.4.) The latter means that the retrial system approaches a classical multiserver multiclass system with *infinite buffer*, provided that the workload process in the orbits tends to infinity. Because the stability/instability of a system is determined by its behavior in a heavily loaded regime, this explains, as we show below, why the stability conditions of both systems coincide.

9.1 Necessary Stability Condition

The classical retrial discipline means that, provided there are n orbital class-i customers, the total retrial rate from orbit i is equal to $n\mu_0^{(i)}$, $i = 1, \ldots, K$. Up to G customers are allowed to queue in the primary system to await service, while the orbits have infinite capacity. This model is illustrated by Fig. 9.1. Denote by $\mathcal{W}(t)$ the remaining amount of work in the *primary system* at instant t, and where, as before

E. Morozov and B. Steyaert, *Stability Analysis of Regenerative Queueing Models*,
https://doi.org/10.1007/978-3-030-82438-9_9

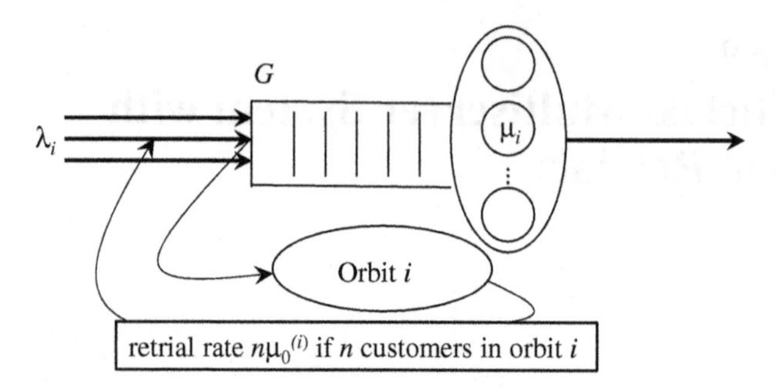

Fig. 9.1 A multiclass m-server finite-buffer retrial system with classical retrials for type-i customers that enter and leave orbit i

$$\rho = \sum_{i=1}^{K} \rho_i = \sum_{i=1}^{K} \lambda_i \mathsf{E} S^{(i)} .$$

Theorem 9.1 *If the system is positive recurrent, then* $\rho < m$ *and the stationary busy probability of an arbitrary server equals* $\mathsf{P}_B = \rho/m$.

Proof Because the proof remains the same for any initial state, we assume a zero initial state and write down the following balance equation which relates $\mathcal{W}(t)$, the total remaining work $\mathsf{W}(t)$ in all orbits, the amount of arrived work $V(t)$, the busy time of all servers $\sum_{i=1}^{m} B_i(t)$ and the total idle time $\widehat{I}(t) = \sum_{i=1}^{m} I_i(t) = tm - \sum_{i=1}^{m} B_i(t)$:

$$V(t) = \sum_{i=1}^{m} B_i(t) + \mathcal{W}(t) + \mathsf{W}(t) = tm - \widehat{I}(t) + \mathcal{W}(t) + \mathsf{W}(t) , \; t \geq 0 , \tag{9.1.1}$$

where, by the positive recurrence,

$$\lim_{t\to\infty} \frac{\mathcal{W}(t)}{t} = \lim_{t\to\infty} \frac{\mathsf{W}(t)}{t} = 0 .$$

As in (8.2.2)–(8.2.4), we obtain from (9.1.1) that, w.p.1,

$$\lim_{t\to\infty} \frac{V(t)}{t} = \rho = \lim_{t\to\infty} \frac{\sum_{i=1}^{m} B_i(t)}{t} = m\mathsf{P}_B = m - \frac{\mathsf{E}\widehat{I_0}}{\mathsf{E}T} < m , \tag{9.1.2}$$

because the mean idle time $\mathsf{E}\widehat{I_0} > 0$ (for instance, see (2.4.7) and (2.2.18)). □

Remark 9.1 The regeneration condition $\min_{1\le i\le K}\mathsf{P}(\tau > S^{(i)}) > 0$ in this system holds automatically because the input process is Poisson.

9.2 Sufficient Stability Condition

Now we show that condition $\rho < m$ is indeed sufficient for the positive recurrence of the zero initial state system.

Theorem 9.2 *If $\rho < m$, then $\mathsf{E}T < \infty$.*

Proof It follows from (9.1.1) that

$$\widehat{I}(t) \ge mt - V(t)\,,\ \ t \ge 0\,.$$

This immediately leads to

$$\liminf_{t\to\infty}\frac{\widehat{I}(t)}{t} \ge m - \rho =: \delta_0 > 0\,. \tag{9.2.1}$$

Denote by $\widehat{I}(t_n) = \widehat{I}_n$. Because, w.p.1,

$$\liminf_{n\to\infty}\frac{\widehat{I}_n}{t_n} \ge \delta_0,\ \ \lim_{n\to\infty}\frac{t_n}{n} = \mathsf{E}\tau\,,$$

then, from (9.2.1), the following lower bound holds w.p.1:

$$\liminf_{n\to\infty}\frac{\widehat{I}_n}{n} = \liminf_{n\to\infty}\frac{\widehat{I}_n}{t_n}\frac{t_n}{n} \ge \delta_0\mathsf{E}\tau =: \varepsilon_0 > 0\,.$$

Now, relying on Fatou's lemma, we obtain

$$\liminf_{n\to\infty}\mathsf{E}\Big(\frac{\widehat{I}_n}{n}\Big) \ge \mathsf{E}\Big(\liminf_{n\to\infty}\frac{\widehat{I}_n}{n}\Big) \ge \varepsilon_0\,. \tag{9.2.2}$$

Denote by $\Delta_n = \widehat{I}_{n+1} - \widehat{I}_n$, the total idle time of the servers in the time interval $[t_n,\, t_{n+1}]$, $n \ge 1$, with $\widehat{I}_1 = 0$. Since $\widehat{I}_n = \sum_{1\le k\le n-1}\Delta_k$, it then follows from (9.2.2) that

$$\liminf_{n\to\infty}\frac{1}{n}\sum_{k=1}^{n-1}\mathsf{E}\Delta_k > 0\,,$$

and there exists a deterministic sequence $n_k \to \infty$ and a constant $d_0 > 0$ such that

$$\inf_k \mathsf{E}\Delta_{n_k} \ge d_0\,. \tag{9.2.3}$$

Next, we introduce an iid sequence

$$\eta_k = \min_{1 \le i \le K} S_k^{(i)}, \quad k \ge 1,$$

of the *shortest* service times, with generic time η and $\mathsf{E}\eta \in (0, \infty)$. Now, for each $n \ge 1$, consider m independent iid sequences $\{\eta_n^{(i)}(k),\ k \ge 1\}$, $i = 1, \ldots, m$, such that all $\eta_n^{(i)}(k) =_{st} \eta$. Then

$$M_i(\tau_n) = \min_{k \ge 1}(k : \eta_n^{(i)}(1) + \cdots + \eta_n^{(i)}(k) \ge \tau_n), \tag{9.2.4}$$

is the number of renewals generated by the sequence $\{\eta_n^{(i)}(k),\ k \ge 1\}$ in the interval $[0,\ \tau_n]$, which are *independent* of τ_n, $i = 1, \ldots, m$. Denote

$$M(\tau_n) = \sum_{i=1}^{m} M_i(\tau_n), \quad n \ge 1, \tag{9.2.5}$$

and note that $\{M(\tau_n)\}$ is the iid sequence with generic element $M(\tau)$. Denote by $\Delta_D(n)$ the total number of customers that leave the system within the interval $[t_n,\ t_{n+1}]$. In order to obtain $M(\tau)$, we use the shortest service times while the real service process contains both idle periods and service times of various classes of customers, and it is therefore easy to see that the following stochastic dominance holds:

$$\Delta_D(n) \le_{st} M(\tau_n) =_{st} M(\tau), \quad n \ge 1. \tag{9.2.6}$$

Because $\mathsf{E}M(\tau) < \infty$ (see Problem 9.1 below), then (uniformly in n)

$$\max_{n \ge 1} \mathsf{P}(\Delta_D(n) > x) \le \mathsf{P}(M(\tau) > x) \to 0, \quad x \to \infty.$$

Denote by $N_n = N(t_n^-)$ the total number of orbital customers at the arrival instant t_n (starting interval τ_n), and define the *minimal retrial rate*,

$$\mu_0 = \min_{1 \le i \le K} \mu_0^{(i)},$$

which corresponds to some class-i_0 customers, i.e., $\mu_0 = \mu_0^{(i_0)}$. In particular, if there are exactly N class-i_0 orbital customers in the system at some time instant, then the remaining retrial time from orbit i_0 at this instant is an exponential random variable with parameter $\mu_0 N$, denoted by $\exp(\mu_0 N)$. Note that if $N_n \ge N$, then $\exp(\mu_0 N)$ is an *upper bound* for the remaining retrial time, at instant t_n, *regardless of the exact allocation* of N_n customers among different orbits. For arbitrary constants N and C_0, we then have

$$\begin{aligned} \mathsf{E}\Delta_n &= \mathsf{E}\big(\Delta_n;\ N_n > N + C_0,\ \Delta_D(n) > C_0\big) && (9.2.7)\\ &+ \mathsf{E}\big(\Delta_n;\ N_n > N + C_0,\ \Delta_D(n) \le C_0\big) && (9.2.8)\\ &+ \mathsf{E}\big(\Delta_n;\ N_n \le N + C_0\big)\,. && (9.2.9) \end{aligned}$$

Using (9.2.6), we may write the term in (9.2.7), for an arbitrary constant $a \ge 0$,

$$\begin{aligned} &\mathsf{E}\big(\Delta_n;\ N_n > N + C_0,\ \Delta_D(n) > C_0\big)\\ &\le \mathsf{E}\big(\Delta_n;\ \Delta_D(n) > C_0\big) \le \mathsf{E}\big(\Delta_n;\ M(\tau_n) > C_0\big)\\ &= \mathsf{E}\big(\Delta_n;\ M(\tau_n) > C_0,\ \tau_n \le a\big) + \mathsf{E}\big(\Delta_n;\ M(\tau_n) > C_0,\ \tau_n > a\big)\,. \end{aligned} \tag{9.2.10}$$

Because τ_n and the set $\{\eta_n^{(i)}(k)\}$ in construction (9.2.4) are mutually independent, we may replace $M(\tau_n)$ by $M(x)$ provided $\tau_n = x$. Since $M(x)$ is non-decreasing and, conditioned on the event $\{\tau_n \le a\}$, the idle time increment Δ_n is upper bounded by ma, this yields

$$\begin{aligned} &\mathsf{E}\big(\Delta_n;\ M(\tau_n) > C_0,\ \tau_n \le a\big)\\ &= \mathsf{E}\big(\Delta_n |\ M(\tau_n) > C_0,\ \tau_n \le a\big)\mathsf{P}(M(\tau_n) > C_0,\ \tau_n \le a)\\ &\le ma\mathsf{P}(M(a) > C_0)\,. \end{aligned} \tag{9.2.11}$$

Conditioned on the event $\{\tau_n = x\}$, denote Δ_n by $\Delta(x)$ and note that $\Delta(x) \le mx$. Then we obtain the following upper bound for the second term in the right-hand side of (9.2.10):

$$\begin{aligned} &\mathsf{E}\big(\Delta_n;\ M(\tau_n) > C_0,\ \tau_n > a\big) \le \mathsf{E}\big(\Delta_n;\ \tau_n > a\big)\\ &\le \int_a^\infty xm\mathsf{P}(\tau \in dx) = me^{-\lambda a}(a + \lambda^{-1})\,. \end{aligned} \tag{9.2.12}$$

Because $\Delta_n \le m\tau_n$, and the interarrival time τ_n and the indicator function $1(N_n \le N + C_0)$ are independent, then the term (9.2.9) is upper bounded by

$$\begin{aligned} \mathsf{E}\big(\Delta_n; N_n \le N + C_0\big) &= \mathsf{E}\big(\Delta_n 1(N_n \le N + C_0)\big)\\ &\le m\mathsf{E}\tau\,\mathsf{P}(N_n \le N + C_0)\,. \end{aligned} \tag{9.2.13}$$

Finally, we need to estimate the term in (9.2.8). Conditioned on the event

$$\mathcal{E}_n := \{N_n > N + C_0,\ \Delta_D(n) \le C_0\}\,, \tag{9.2.14}$$

at most C_0 customers leave the system in the time interval $[t_n,\ t_{n+1}]$, and after each departure the total orbit size is still no less than N. For this reason, the mean idle period (of any server) after each departure is upper bounded by

$$\mathsf{E}\exp(\mu_0 N) = \frac{1}{N\mu_0}. \tag{9.2.15}$$

Note that within the interval $[t_n, t_{n+1}]$ there exist at most m initial idle periods of the servers (at instant t_n) and, conditioned on the event $\mathcal{E}_n$, at most C_0 idle periods after the departures. Together with (9.2.15), this leads to the following upper bound for the term (9.2.8):

$$\mathsf{E}(\Delta_n 1(\mathcal{E}_n)) = \mathsf{E}(\Delta_n|\mathcal{E}_n)\mathsf{P}(\mathcal{E}_n) \le \mathsf{E}(\Delta_n|\mathcal{E}_n) \le \frac{m + C_0}{N\mu_0}. \tag{9.2.16}$$

We now return to inequality (9.2.3), and choose a in (9.2.12) such that

$$me^{-\lambda a}(a + \lambda^{-1}) \le \frac{d_0}{8}, \tag{9.2.17}$$

and then, because of $M(a) < \infty$ w.p.1, choose C_0 such that

$$ma\mathsf{P}(M(a) > C_0) \le \frac{d_0}{8}, \tag{9.2.18}$$

and, finally, choose N large enough such that

$$\frac{m + C_0}{N\mu_0} \le \frac{d_0}{8}. \tag{9.2.19}$$

Now we assume that the total orbit size satisfies

$$N_n \Rightarrow \infty, \ n \to \infty. \tag{9.2.20}$$

Then in particular, for each N and C_0, this leads to

$$\lim_{n\to\infty} \mathsf{P}(N_n \le N + C_0) = 0,$$

and we can select n_0 large enough such that, for all $n \ge n_0$,

$$m\,\mathsf{E}\mathcal{T}\,\mathsf{P}(N_n \le N + C_0) \le \frac{d_0}{8}. \tag{9.2.21}$$

Therefore, combining the upper bounds (9.2.17)–(9.2.19) and (9.2.21), we obtain from (9.2.7)–(9.2.16) that for all $n \ge n_0$

$$\mathsf{E}\Delta_n \le \frac{d_0}{2},$$

contradicting (9.2.3). Consequently, assumption (9.2.20) is false, and there exist a deterministic sequence $n_k \to \infty$, and constants $\delta > 0,\ D \geq 0$, such that

$$\inf_k \mathsf{P}(N_{n_k} \leq D) \geq \delta\,. \tag{9.2.22}$$

Let Q_n be the number of customers in the primary system at instant t_n^-, then, because of $Q_n \leq G+m$ w.p.1,

$$\inf_k \mathsf{P}\big(N_{n_k} \leq D,\ Q_{n_k} \leq m+G\big) \geq \delta\,.$$

In addition, since the primary system is finite, then the remaining work sequence $\{\mathcal{W}(t_n^-) =: \mathcal{W}_n\}$ is tight, and for an appropriate constant D_0, we may write

$$\mathsf{P}\big(N_{n_k} \leq D,\ Q_{n_k} \leq m+G,\ \mathcal{W}_{n_k} \leq D_0\big) \geq \frac{\delta}{2}\,. \tag{9.2.23}$$

Recall that $\{n_k\}$ is a deterministic subsequence such that $\{t_{n_k}\} \subseteq \{t_n\}$. The rest of proof is based on the possibility to serve each orbital customer within a finite time interval. Then, relying on the unboundedness of τ, we can unload the system within one interarrival time, i.e., in the interval $[t_{n_k},\ t_{n_k+1}]$, with a positive probability. This leads to

$$\inf_k \mathsf{P}(\theta(n_k) = 1) > 0\,,$$

and hence, $\theta(n_k) \not\Rightarrow \infty$ as $k \to \infty$, implying $\mathsf{E}\theta < \infty$ and $\mathsf{E}T < \infty$. □

For more details of the proof and the more challenging case with renewal input and finite τ, we refer to the paper [1].

Problem 9.1 Prove that $\mathsf{E}M(\tau) < \infty$ using (9.2.5), representation

$$\mathsf{E}M_i(\tau) = \lambda \int_0^\infty \mathsf{E}M_i(x)e^{-\lambda x}dx\,,$$

and the elementary renewal theorem

$$\lim_{x\to\infty} \frac{\mathsf{E}M_i(x)}{x} = \frac{1}{\mathsf{E}\eta},\quad i = 1,\ldots,K\,.$$

Problem 9.2 Prove inequality (9.2.12) in detail.

9.3 Notes

The analysis in this chapter has been inspired by paper [3]. An important study presented in [4] contains both the stability conditions and the bounds on the rate of convergence to stationarity for a single-server non-Markovian retrial system, also with non-exponential retrials. The classical retrial discipline is a particular case of a much more general *linear repeated requests strategy*, see for instance, the fundamental monograph [5] on this topic. An extension of the results presented above to the system with the New-Better–Than–Used (NBU) retrial times is given in paper [6]. The stability analysis of a single-class $GI/G/m/0$-type retrial queue has been developed in [2]. The tightness of the processes in a finite queuieng system is discussed in [7], which in turn is partly based on the results from [8].

References

1. Morozov, E., Phung-Duc, T.: Regenerative analysis of two-way communication orbit-queue with general service time. In: Proceedings International Conference Queueing Theory and Network Applications (QTNA 2018, LNCS 10932), pp. 22–32. Tsukuba (2018)
2. Morozov, E.: A multiserver retrial queue: Regenerative stability analysis. Queueing Syst. **56**, 157–168 (2007)
3. Morozov, E., Phung-Duc, T.: Stability analysis of a multiclass retrial system with classical retrial policy. Perf. Eval. **112**, 15–26 (2017)
4. Altman, E., Borovkov, A.A.: On the stability of retrial queues. Queueing Syst. **26**, 343–363 (1997)
5. Artalejo, J.R., Gómez-Corall, A.: Retrial Queueing Systems: A Computational Approach. Springer, Berlin (2008)
6. Morozov, E., Nekrasova, R.: Stability conditions of a multiclass system with NBU retrials. In: Proceedings 14th International Conference on Queueing Theory and Network Applications (QTNA 2019, LNCS 11688), pp. 51–63. Ghent (2019)
7. Morozov, E.: The tightness in the ergodic analysis of regenerative queueing processes. Queueing Syst. **27**, 179–203 (1997)
8. Shanbhag, D.N.: Some extentions of Takacs's limit theorems. J. Appl. Probab. **11**, 752–761 (1974)

Chapter 10
Other Related Models

Evidently, many more applications of the regenerative stability analysis technique can be considered. We address just some of these applications in this chapter, by respectively focusing on the stability analysis of an optical buffer model, on discrete-time queues, and on regenerative networks. We discuss only the new aspects of the related stability analysis, by comparing them with the ones that have been presented in the preceding chapters, and by highlighting the similarities.

10.1 An Optical Buffer System

First we outline the stability analysis of a system with optical buffers. As we will observe, this model is an interesting combination of the previously studied buffered multiserver system (Chap. 2), a state-dependent system (Chap. 5) and a retrial system (Chap. 7).

10.1.1 Description of the Model

Since light can not be physically stored, in an optical buffer system contention between incoming *optical bursts* (i.e., customers) is resolved by means of a *fiber delay line* (FDL) buffer, which can delay, if necessary, optical bursts (or packets) until the designated channel becomes available again. We consider an optical buffer with $m \geq 1$ *wavelengths* (servers). (The single-wavelength system was analyzed in [1].) Denote by t_k the arrival instant of the kth optical burst, $\tau_k = t_{k+1} - t_k$ as the kth interarrival time (with generic element τ), S_k as the *burst size (transmission or service time)* of burst k (with generic element S), and let $U_k = S_k - \tau_k$, $k \geq 1$ (with

E. Morozov and B. Steyaert, *Stability Analysis of Regenerative Queueing Models*,
https://doi.org/10.1007/978-3-030-82438-9_10

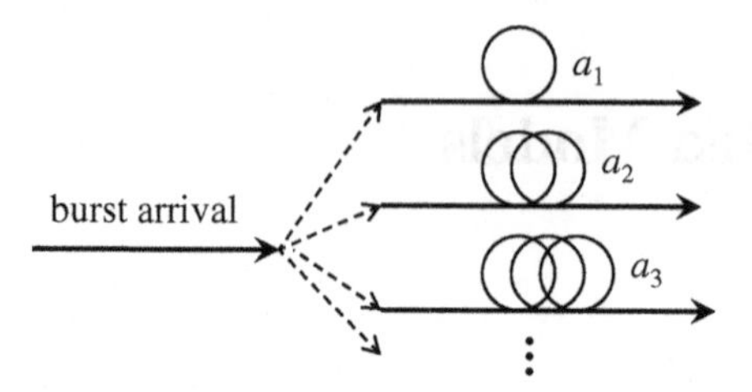

Fig. 10.1 An optical buffer system

generic element U). By assumption, $\{\tau_k\}$ and $\{S_k\}$ are independent iid sequences. The system is characterized by a set $\mathcal{A} = \{a_1, a_2, \ldots\}$ of available FDL lengths or *delays*, such that

$$0 < a_1 < a_2 < \ldots, \quad \lim_{i \to \infty} a_i = \infty ,$$

implying that there is always a sufficiently long delay line available for a newly arriving optical burst, see Fig. 10.1 (Such a model is used as a reference model, for instance, in [2, 3].) Note that the FDL lengths indeed represent the *time* needed to traverse the optical coil of the corresponding FDL. The main distinctive feature of the optical buffer system is the constraint imposed by the FDLs to maintain the FCFS service discipline and at the same time avoid overlap with previous bursts. The optical buffers in general cannot assign the exact required delay x, but only a delay satisfying condition $a_{i-1} < x \leq a_i$, which can be defined as

$$\lceil x \rceil_{\mathcal{A}} := \min(a_i \in \mathcal{A} : a_i \geq x) , \; x > 0 .$$

The difference

$$\Delta(x) := \lceil x \rceil_{\mathcal{A}} - x \geq 0 ,$$

the time during which the wavelength remains unused, is called a *void* (or gap) and represents the amount of *capacity loss*, provided the delay equals x. This gap implies that the service discipline is *not non-idling* (in this regard, also see the last comment in Sect. 7.7) and is rather similar to the delay in a retrial system where a server is not occupied immediately after it becomes free but has to wait for either a successful attempt of an orbital customer or a new customer arrival.

At any instant t, we define the *scheduling horizon* of a wavelength as the amount of time, starting from instant t, until all bursts assigned to the wavelength have left this wavelength, provided that no new arrivals would occur after time instant t. Evidently, this concept bears analogues to the remaining work for a given server in a multiserver buffered system. Also, let $H_k^{(i)}$ denote the ith *smallest* scheduling horizon upon arrival of the k-h burst, i.e., at time instant t_k. In other words, $H_k^{(i)}$ is the earliest time, measured from t_k on, when the ith wavelength becomes free of all previously arrived (and scheduled) bursts. Note that in this arrangement, the index i does not refer to wavelength i, but to the order by which wavelengths become free of all bursts assigned to them. (If wavelength i has no bursts assigned to it, then

$H_k^{(i)} = 0$.) Next, we introduce the m-dimensional Markov process

$$\mathbf{H}_k = (H_k^{(1)}, \dots, H_k^{(m)}), \;\; H_k^{(1)} \le \cdots \le H_k^{(m)} ,$$

which, in view of the previous definitions and description of an FDL-based optical buffer's operation, satisfies the following Kiefer–Wolfowitz-type recursion (see also (2.5.27)):

$$\mathbf{H}_{k+1} = R\left(\left(\left\lceil H_k^{(1)} \right\rceil_{\mathcal{A}} + U_k\right)^+, \left(H_k^{(2)} - \tau_k\right)^+, \dots, \left(H_k^{(m)} - \tau_k\right)^+\right), \; k \ge 1 ,$$

where, as before, the operator R rearranges the components in an increasing order.

Problem 10.1 Show that $\{\mathbf{H}_k\}$ is a Markov chain.

Since $H_k^{(m)} = 0$ implies $H_k^{(i)} = 0, \; i = 1, \dots, m-1$, then we can define the regeneration instants of the process $\{\mathbf{H}_k\}$ as follows:

$$T_{n+1} = \inf_{k \ge 1} \left(t_k > T_n : H_k^{(m)} = 0\right) , \;\; n \ge 0 \;\; (T_0 := 0) .$$

10.1.2 Stability Analysis

Denote by $g_n = a_{n+1} - a_n$, and let

$$\Delta^* = \sup_{n \ge 0} g_n \; ; \quad \delta^* = \inf_{n \ge 0} g_n \; ; \quad \Delta_\infty = \limsup_{n \to \infty} g_n \; ,$$

assuming $a_0 = 0$, $\delta^* > 0$ and $\Delta^* < \infty$. In particular, Δ_∞ is the distance between two consecutive 'infinitely long' FDLs, and hence

$$\limsup_{x \to \infty} \Delta(x) \le \Delta_\infty \; ,$$

as any gap size is always upper bounded by the difference between the corresponding adjoining available delay values in the FDL set. Since $\mathsf{E}S < \infty$, $\mathsf{E}\tau < \infty$, then

$$\mathsf{E}U \in (-\infty, +\infty) . \tag{10.1.1}$$

Let

$$\Delta_H(k) = \sum_{i=1}^{m} (H_{k+1}^{(i)} - H_k^{(i)}) , \;\; k \ge 1 ,$$

and define

$$\mathbf{y} = (y_1, \ldots, y_m), \; y_1 \le \cdots \le y_m \, .$$

Consider the (conditional) mean increment, satisfying

$$\begin{aligned}\mathsf{E}(\Delta_H(k)|\mathbf{H}_k = \mathbf{y}) &= \mathsf{E}\Big[(\lceil y_1 \rceil_{\mathcal{A}} + U_k)^+ - y_1\Big] + \mathsf{E}\sum_{i=2}^{m} \big[(y_i - \tau_k)^+ - y_i\big] \\ &= -y_1 \mathsf{P}(U \le -y_1 - \Delta(y_1) + \mathsf{E}\big[(U + \Delta(y_1))\mathbf{1}(U > -y_1 - \Delta(y_1))\big]) \\ &\quad + \mathsf{E}\sum_{i=2}^{m} \big[(y_i - \tau_k)^+ - y_i\big] \, ,\end{aligned}$$

which in view of the Markov property does not depend on index k and further transforms into

$$\begin{aligned}\mathsf{E}(\Delta_H(k)|\mathbf{H}_k = \mathbf{y}) &= -y_1 \mathsf{P}\Big(U \le -y_1 - \Delta(y_1)\Big) + \Delta(y_1)\mathsf{P}\Big(U > -y_1 - \Delta(y_1)\Big) \\ &\quad + \int_{z \ge -y_1 - \Delta(y_1)} z F_U(dz) + \mathsf{E}\sum_{i=2}^{m} \big[(y_i - \tau_k)^+ - y_i\big] \, , \qquad (10.1.2)\end{aligned}$$

where F_U represents the distribution function of U. Since $\mathsf{E}\tau < \infty$, then it is easy to check that, as $y \to \infty$,

$$\mathsf{E}(y - \tau)^+ - y) = -\mathsf{E}(\tau; \tau \le y) - y\mathsf{P}(\tau > y) \to -\mathsf{E}\tau \, ,$$

implying (because $y_1 = \min_i y_i$)

$$\lim_{y_1 \to \infty} \mathsf{E}\sum_{i=2}^{m} \big[(y_i - \tau_k)^+ - y_i\big] = -(m-1)\mathsf{E}\tau \, . \qquad (10.1.3)$$

On the other hand, in view of (10.1.1), as $y_1 \to \infty$,

$$\begin{aligned}\mathsf{P}(U > -y_1 - \Delta(y_1)) &\ge \mathsf{P}(U > -y_1) \to 1 \, , \\ 0 \ge -y_1 \mathsf{P}(U \le -y_1 - \Delta(y_1)) &\ge -y_1 \mathsf{P}(U \le -y_1) \to 0 \, . \qquad (10.1.4)\end{aligned}$$

Then it follows that

$$\limsup_{y_1 \to \infty} \Delta(y_1)\mathsf{P}(U > -y_1 - \Delta(y_1)) = \limsup_{y_1 \to \infty} \Delta(y_1) \le \Delta_\infty \, , \qquad (10.1.5)$$

and

$$\lim_{y_1\to\infty}\int_{z\geq -y_1-\Delta(y_1)} zF_U(dz) = \mathsf{E}U - \lim_{y_1\to\infty}\int_{-\infty}^{-y_1-\Delta(y_1)} zF_U(dz) = \mathsf{E}U\,. \quad (10.1.6)$$

Then, we obtain from (10.1.2)–(10.1.6) the following inequality:

$$\limsup_{y_1\to\infty} \mathsf{E}(\Delta_H(k)|\mathbf{H}_k = \mathbf{y}) \leq \Delta_\infty + \mathsf{E}S - m\mathsf{E}\tau\,. \quad (10.1.7)$$

Now, with a minor change of the proof used in the analysis of the classical multiserver system in Chap. 2, one can establish the tightness of the sequence $\Delta_H^{(k)} := H_k^{(m)} - H_k^{(1)}$, $k \geq 1$. The main stability result is then as follows:

Theorem 10.1 *Assume that conditions*

$$P(\tau > \Delta^* + S) > 0\,, \quad (10.1.8)$$

$$\Delta_\infty + \mathsf{E}S < m\mathsf{E}\tau\,,$$

hold. Then $\mathsf{E}T < \infty$.

The proof of Theorem 10.1 is quite standard: we assume that the *minimal* component satisfies

$$H_k^{(1)} \Rightarrow \infty\ ,\quad k \to \infty\,, \quad (10.1.9)$$

and then apply the asymptotic result (10.1.7) to obtain a contradiction to (10.1.9). Next, we rely on the tightness of the sequence $\{\Delta_H^{(k)}\}$, implying $H_k^{(m)} \not\Rightarrow \infty$ as well, and then use the regeneration assumption (10.1.8) to unload the system and show that the process $\{\mathbf{H}_k\}$ regenerates with a positive probability in a finite time. This implies $\mathsf{E}T < \infty$. (For additional details concerning this analysis, we refer to [1].)

Problem 10.2 Using $\mathsf{E}\tau < \infty$ show that $\lim_{x\to\infty} x\mathsf{P}(\tau > x) = 0$. Check (10.1.2), and prove (10.1.4).

10.2 A Discrete-Time System with Renewal-Type Interruptions

In this section we outline the stability analysis of a discrete-time multiserver system with *renewal-type server interruptions* based on results obtained in [4]. An indispensable new element of the analysis is the *synchronisation* of the independent interruption processes, with the purpose of constructing common regeneration instants by exploiting the discrete structure of the distributions.

10.2.1 Description of the Model

We consider an m-server discrete-time queueing system with renewal input with arrival instants $\{t_k\}$, an *aperiodic* generic interarrival time τ with rate $\lambda = 1/\mathsf{E}\tau$ and with iid service times $\{S_n\}$, in which the servers are assumed to be *unavailable (blocked)* from time-to-time. We denote by $X_n^{(i)}$ and $Y_n^{(i)}$ the lengths of the nth blocked and nth available period of the ith server, respectively, $i = 1, \dots, m$. For each i, the pairs $(X_n^{(i)}, Y_n^{(i)})$, $n \geq 1$, are assumed to form an iid sequence. Moreover, we assume that $X_n^{(i)}$ and $Y_n^{(i)}$ are independent as well, although a dependence between these random variables is possible in the framework of our analysis and is natural, for instance, in the context of $GI/G/1$ *preemptive priority queues*, see [4]. The length of the nth *cycle* of server i (which consists of a blocked period followed by an available period) is given by

$$Z_n^{(i)} := X_n^{(i)} + Y_n^{(i)}, \quad i = 1, \dots, m .$$

Note that $\{Z_n^{(i)}, \; n \geq 1\}$ is an iid sequence, with generic element $Z^{(i)}$. We assume that $Z^{(i)}$ is aperiodic, and denote by $Z_i(t)$ the remaining renewal time in the (renewal) process generated by the sequence $\{Z_n^{(i)}\}$.

An unavailability period can interrupt a customer service, and we adopt either the *preemptive-resume* or the *preemptive-repeat* service discipline. In the former case, service continues after an interruption, whereas a new (and independent) service time starts after an interruption in the latter case. In both cases, customers remain with the same server until service completion.

Denote by $\nu(t)$ the queue size, which includes the departures but excludes arrivals at instant t (if any). For simplicity, we also assume a zero initial state $t_1 = \nu(0) = 0 = Z_i(0)$ (for all i). Denote by $\tau(t)$ the remaining interarrival time at instant t and let $\mathcal{Z}_n^{(i)} = \sum_{k=1}^{n} Z_k^{(i)}$, $n \geq 1$, be the completion instants of the cycles of server i. We note that, by construction,

$$\tau(t_k) = 0\,, \; k \geq 1 \quad \text{and} \quad Z_i(\mathcal{Z}_n^{(i)}) = 0\,, \; n \geq 1\,, \; i = 1, \dots, m\,. \tag{10.2.1}$$

Then $T_0 = 0$ and

$$T_{n+1} = \inf_{t \geq 1} \big(t > T_n : Z_1(t) = \cdots = Z_m(t) = \tau(t) = \nu(t) = 0\big), \;\; n \geq 0\,, \tag{10.2.2}$$

represent the regeneration instants of the system, with generic period T. It easily follows from (10.2.1) that, for such instants to exist, the regeneration points $\{T_n\}$ must be a subsequence of the arrival instants and server cycle completion instants,

$$\{T_n, \; n \geq 1\} \subseteq \{t_k, \; k \geq 1\} \bigcap \bigcap_{i=1}^{m} \{\mathcal{Z}_n^{(i)}, \; n \geq 1\}\,. \tag{10.2.3}$$

For this reason, we assume that the following condition holds: there exist integers $a_1, \ldots, a_m$ such that

$$\min_{1\le i\le m} \mathsf{P}(\tau = a_i Z^{(i)}) > 0\,, \tag{10.2.4}$$

and

$$\min_{1\le i\le m} \mathsf{P}(Y^{(i)} > S) > 0\,. \tag{10.2.5}$$

Both assumptions (10.2.4) and (10.2.5) are used to *synchronize* the renewal points $\{t_k\}$ of the input process and the renewal points $\{\mathcal{Z}_n^{(i)}\}$ of the interruption processes during the unloading of the system that results in a new regeneration point, being a common one both for the input process and for all processes $\{Z_n^{(i)}\}$ [4]. Note that conditions (10.2.4) and (10.2.5) hold for a wide class of discrete distributions, for instance, if each $Y^{(i)}$ has infinite support, and the support of τ equals a set of positive integers. Due to the aperiodicity assumption and $\mathsf{E}Z^{(i)} < \infty$, $\mathsf{E}\tau < \infty$, the following limits exist:

$$\lim_{t\to\infty} \frac{1}{t}\sum_{k=1}^{t} \mathsf{1}(\tau(k) = 0) = \lim_{t\to\infty} \mathsf{P}(\tau(t) = 0) = \frac{1}{\mathsf{E}\tau} = \lambda\,,$$

$$\begin{aligned}\lim_{t\to\infty} \frac{1}{t}\sum_{k=1}^{t} \mathsf{1}(Z_i(k) = 0) &= \lim_{t\to\infty} \mathsf{P}(Z_i(t) = 0)\\ &= \frac{1}{\mathsf{E}Z^{(i)}} =: \lambda_0^{(i)},\ i = 1, \ldots, m\,.\end{aligned} \tag{10.2.6}$$

10.2.2 Stability Analysis

Denote by $\lambda_0 = \sum_{1\le i\le m} \lambda_0^{(i)}$, and outline the proof of the following statement.

Theorem 10.2 *Assume that, in case of a preemptive-repeat discipline, conditions (10.2.4), (10.2.5) and the negative drift condition*

$$(\lambda + \lambda_0)\mathsf{E}S + \sum_{i=1}^{m} \lambda_0^{(i)}\mathsf{E}X^{(i)} < m\,, \tag{10.2.7}$$

hold. Then $\mathsf{E}T < \infty$.

Proof Denote by

$$I(t) = \sum_{n=1}^{t} \mathsf{1}(\nu(n) < m)\,,\ t \ge 0\,.$$

Also, the total amount of blocked time in the interval $[0, t]$ is given by $X(t) = \sum_{i=1}^{m} X_i(t)$, where $X_i(t)$ is the blocked time of server i. Then the work $V(t)$ that has arrived within interval $[0, t]$ satisfies the inequality

$$V(t) \geq mt - mI(t) - X(t), \; t \geq 0\,, \tag{10.2.8}$$

where we have ignored the remaining work (see (2.2.5)). Again, $\{S_n\}$ represents the sequence of original iid service times, and let $A(t)$ be the number of arrivals in the interval $[0, t]$. Also define $\{S_n^{(i)}\}$ as the iid service times that are assigned after the interruptions of server i, which are distributed as a generic service time S. Since there is at most one service interruption for every blocked period, we have that

$$V(t) \leq \sum_{j=1}^{A(t)} S_j + \sum_{i=1}^{m} \sum_{n=1}^{G_i(t)} S_n^{(i)} =: \mathbb{V}(t)\,, \tag{10.2.9}$$

where $G_i(t)$ is the number of breakdowns of server i within the interval $[0, t]$. In view of (10.2.6), renewal process $\{G_i(t)\}$ satisfies

$$\lim_{t\to\infty} \frac{G_i(t)}{t} = \frac{1}{\mathsf{E}Z^{(i)}} = \lambda_i^{(0)}\,.$$

Then, again by invoking the SLLN, this leads to

$$\lim_{t\to\infty} \frac{1}{t}\mathbb{V}(t) = \lambda \mathsf{E}S + \sum_{i=1}^{m} \lambda_0^{(i)} \mathsf{E}S = (\lambda + \lambda_0)\mathsf{E}S\,. \tag{10.2.10}$$

Since the blocked time $\{X_i(t)\}$ is a positive recurrent cumulative process with regeneration period $Z^{(i)}$ and 'cycle' increment $X^{(i)}$, then

$$\lim_{t\to\infty} \frac{X_i(t)}{t} = \frac{\mathsf{E}X^{(i)}}{\mathsf{E}Z^{(i)}} = \lambda_0^{(i)}\mathsf{E}X^{(i)}, \; i = 1, \ldots, m\,. \tag{10.2.11}$$

Because, due to (10.2.8) and (10.2.9),

$$I(t) \geq t - \frac{1}{m}(\mathbb{V}(t) + X(t))\,,$$

then (10.2.7), (10.2.10) and (10.2.11) imply

$$\liminf_{t\to\infty} \frac{I(t)}{t} \geq 1 - \frac{(\lambda + \lambda_0)\mathsf{E}S}{m} - \frac{\sum_{i=1}^{m} \lambda_0^{(i)} \mathsf{E}X^{(i)}}{m} > 0\,,$$

and, by invoking Fatou's lemma, this yields

$$\liminf_{t\to\infty} \frac{1}{t} \mathsf{E} I(t) = \liminf_{t\to\infty} \frac{1}{t} \sum_{1\le n\le t} \mathsf{P}(\nu(n) < m) > 0 .$$

Hence, for a (deterministic) subsequence $n_k \to \infty$ and some $\varepsilon > 0$, we obtain

$$\inf_k \mathsf{P}(\nu(n_k) < m) \ge \varepsilon . \tag{10.2.12}$$

Next, we define the process that jointly counts the remaining times of the input process and all renewal breakdown processes, i.e., the process

$$\mathbf{U}(t) = \{\tau(t),\ Z_i(t),\ i = 1, \dots, m\},\ t \ge 0 .$$

The common renewal points of this process are then defined as

$$\mathbb{T}_0 = 0,\ \mathbb{T}_{n+1} = \min_{n\ge 0}(t > \mathbb{T}_n : \mathbf{U}(t) = \mathbf{0}) ,$$

with generic renewal period length $\mathbb{T}$. (Consequently, the set $\{\mathbb{T}_n\}$ constitutes a subset of the arrival instants and server cycle completion instants, see also (10.2.3).) Also, let $\mathbb{T}(t)$ be the remaining renewal time process at instant t. Then, using the independence of the input process and breakdown processes, and also relying on (10.2.6), we find that

$$\lim_{t\to\infty} \mathsf{P}(\mathbb{T}(t) = 0) = \frac{1}{\mathsf{E}\mathbb{T}} = \lim_{t\to\infty} \mathsf{P}\big(\mathbf{U}(t) = \mathbf{0}\big) = \lambda \prod_{i=1}^{m} \lambda_0^{(i)} > 0 .$$

Consequently, the renewal process $\{\mathbb{T}_n\}$ is positive recurrent, and hence the process $\{\mathbb{T}(t)\}$ is tight. Then it follows from (10.2.12) that

$$\inf_k \mathsf{P}\big(\nu(n_k) < m,\ \mathbb{T}(n_k) \le C\big) \ge \frac{\varepsilon}{2} ,$$

for some constant C. (Note that the remaining amount of work at the series of time instants n_k is a tight process, which can be shown by adopting the same arguments that have been repeatedly used in a continuous-time setting for the deterministic sequence of time instants $\{z_i\}$.) The rest of the proof is straightforward and based on assumptions (10.2.4) and (10.2.5), which allow us to show in a standard way that the remaining regeneration time $T(t)$ associated with regenerations (10.2.2) does not go to infinity. □

Remark 10.1 The negative drift condition (10.2.7) has a nice qualitative explanation. The term $\lambda \mathsf{E}S$ relates to the incoming workload, the term $\sum_{i=1}^{m} \lambda_0^{(i)} \mathsf{E}X^{(i)}$ describes the loss of capacity caused by the interruptions, and the term $\lambda_0 \mathsf{E}S$ expresses the loss caused by the service repetitions of interrupted customers.

In case of *preemptive-resume* interruptions, the negative drift condition takes into account the blocked periods, but the interruptions bring no additional workload into the system. A slight adaptation of the above proof leads to the following result:

Theorem 10.3 *The statement of Theorem 10.2 holds for a system with preemptive-resume service interruptions provided that assumption (10.2.7) is replaced by*

$$\sum_{i=1}^{m} \lambda_0^{(i)} \mathsf{E}X^{(i)} + \lambda \mathsf{E}S < m .$$

Remark 10.2 Although the aperiodicity assumption is sufficient to enable the synchronization of the renewal points of the input and breakdown processes, in order to unload the system we need more specific conditions, such as (10.2.4) and (10.2.5) (which in turn can be replaced by alternative conditions [4]). These conditions allow to maintain the synchronization at each step of the unloading process.

10.3 Regeneration in Queueing Networks

As we mentioned before, the regenerative analysis of *queueing networks* is in general beyond the scope of this book. Nevertheless, in this section we highlight how to prove the *necessary* stability conditions of some regenerative queueing networks using the approach that we have presented in previous chapters.

Open networks. First, we consider a particular case of a single-class *open generalized Jackson network* with zero initial state, N infinite-capacity single-server stations, a unique renewal input process with a generic interarrival time τ and rate $\lambda = 1/\mathsf{E}\tau$. (The *classical Jackson network* is described by exponential distributions, see [5, 6].) The routes of customers inside the network are governed by a $N \times N$ *routing matrix* where P_{ij} is the probability that a customer leaving station i is routed to station j. It is assumed that there are no isolated stations, and each customer eventually leaves the network. Such a network regenerates when a newly arriving customer encounters a *fully idle network*, if such a state exists. (This concept is quite similar to the definition of regenerations in the queueing systems considered throughout the present book.) Denote by $a_i(t)$ be the number of visits to station i generated by all $A(t)$ customers arriving in the network during the time interval $[0, t]$, and let $a_n^{(i)}$ be the number of visits to station i by customer n. (We note that $\{a_n^{(i)},\ n \geq 1\}$ are iid for each i.) Also denote by $N_i(t)$ the number of external arrivals that are *directly routed* (with a probability P_{0i}) to station i during $[0, t]$. Define the indicator function $1_{ij}^{(n)} = 1$ if and only if the nth customer leaving station i is directed to station j. It is assumed that these indicators are independent of the states of the stations. (This type of routing is also called *Bernoulli*, see for instance [7].) The following relations are now evident:

$$a_i(t) = \sum_{n=1}^{A(t)} a_n^{(i)} = N_i(t) + \sum_{j=1}^{N} \sum_{n=1}^{a_j(t)} 1_{ji}^{(n)}, \; i = 1, \dots, N; \; t \geq 0 \,. \quad (10.3.1)$$

In view of the properties of the network, $\max_i \mathsf{E}a^{(i)} < \infty$, and the following *potential inflow* rate for each station i exists:

$$\lambda_i := \lim_{t \to \infty} a_i(t)/t \,, \quad (10.3.2)$$

and we obtain from (10.3.1) the well-known *traffic equations* relating the unknown rates $\{\lambda_i\}$:

$$\lambda_i = \lambda P_{0i} + \sum_{j=1}^{N} \lambda_j P_{ji}, \; i = 1, \dots, N \,. \quad (10.3.3)$$

Problem 10.3 Using the SLLN, derive (10.3.3) from (10.3.1) and (10.3.2).

Denote by $\{S_n^{(i)}\}$ the iid service times at station i, and let $\lambda_i \mathsf{E}S^{(i)} = \rho_i$. It has been proved in [8] that, provided that the interarrival time τ is unbounded, then condition

$$\max_{1 \leq i \leq N} \rho_i < 1 \,, \quad (10.3.4)$$

implies the positive recurrence of this network. (Under assumption (10.3.4), the stability of a general open network was established in paper [9], where the analysis does not refer to the regenerative method.)

We note that if condition (10.3.4) is violated for some station i_0, i.e., $\rho_{i_0} \geq 1$, then we can not assert in general that station i_0 is 'unstable', as the following simple example shows [10]. Consider a tandem network with two single-server stations, a Poisson input with rate λ to station 1, iid service times with rate μ_i at station $i = 1, 2$, and with routing probabilities $P_{01} = P_{12} = P_{20} = 1$. (With probability P_{20} any customer served by station 2 leaves the network, and index 0 represents the 'outside world'.) Assume that

$$\lambda > \mu_2 > \mu_1 \,,$$

then the solution of (10.3.3), $\lambda_1 = \lambda_2 = \lambda$, results in $\rho_i > 1$ for both stations. However, as it is easy to see, the queue size (and workload process) at station 2 remains tight because the real (long-range) input rate in station 2 is μ_1, implying the traffic intensity $\mu_1/\mu_2 < 1$. An evident reason for this effect is that the rates λ_i satisfying (10.3.3) are 'potential', and in our example station 2 will not receive the full amount of work that is routed to it because station 1 is the *bottleneck* [11] and accumulates the excess load intended for station 2.

As we mentioned above, an evident regeneration condition in the network is the unboundedness of the interarrival time τ. However, classical regenerations based on the empty state of the network can be constructed under a weaker assumption, namely that there exists a *route* $\mathcal{R}$ such that an external customer can cross the network along the route $\mathcal{R}$ with a positive probability, and moreover, that

$$\mathsf{P}(\tau > \sum_{i \in \mathcal{R}} S^{(i)}) > 0 , \tag{10.3.5}$$

holds, implying that with a positive probability an external customer can cross the *empty stations* of $\mathcal{R}$ *before* the next external customer arrives. It is easy to recognize (10.3.5) as a regeneration condition. (Indeed, conditions (10.3.5) and (10.3.4) imply the existence of the positive recurrent process of classical regenerations in a wide class of generalized Jackson networks [12–14].) In a positive recurrent regenerative network the inflow to each station i also is positive recurrent regenerative with rate λ_i defined in (10.3.2), and the equation

$$V_i(t) = W_i(t) + t - I_i(t) , \tag{10.3.6}$$

relates the amount of the arrived work $V_i(t)$, the remaining work $W_i(t)$, and the idle time $I_i(t)$. It follows from (10.3.2) that, for each i,

$$\lim_{t\to\infty} \frac{V_i(t)}{t} = \lim_{t\to\infty} \frac{\sum_{n=1}^{a_i(t)} S_n^{(i)}}{a_i(t)} \frac{a_i(t)}{t} = \lambda_i \mathsf{E} S^{(i)} = \rho_i . \tag{10.3.7}$$

On the other hand, by the positive recurrence of the involved processes, the stationary probability π_0 that the network is completely idle is positive, and because

$$\lim_{t\to\infty} \frac{I_i(t)}{t} \geq \lim_{t\to\infty} \frac{1}{t} \int_0^t 1\big(\sum_{i=1}^{N} W_i(u) = 0\big) du =: \pi_0 > 0 , \tag{10.3.8}$$

then it follows from (10.3.6) that $\rho_i < 1$. Thus condition (10.3.4) is indeed necessary for positive recurrence of the network.

Problem 10.4 Explain why, in the positive recurrent network, in order to prove $V_i(t)/t \to \rho_i$ in (10.3.7), we may use the *potential* number $a_i(t)$ of visits to station i instead of the actual number of visits in $[0, t]$.

Closed networks. In a *closed network*, a finite number of customers circulate between N single-server stations, and the total (i.e., aggregated over all stations) work is a tight process. In particular, if there exists a route $\mathcal{R}$ from station i to station i such that

$$\mathsf{P}(S^{(i)} > \sum_{j \in \mathcal{R},\, j \neq i} S^{(j)}) > 0 ,$$

then there exists a positive recurrent process of *classical* regenerations for the vector queue-size process, which occurs when *all customers are collected in station i and the customer at the front ends his service* [8, 13, 15, 16].

Multiclass networks. In a *multiclass* queueing network, the routing matrix describes the transitions *between classes*, although a one-to-one correspondence between classes and stations exists as well [6], and this allows to formulate regeneration conditions similar to (10.3.5) to apply regenerative stability analysis [14].

10.4 Notes

The material and terminology in Sect. 10.1 are borrowed from the papers [1, 17]. Both *optical burst switching* [18, 19] and *optical packet switching* [20, 21] increase bandwidths by avoiding the conversion from light to electricity but require optical buffering. Also see the related papers [22–24]. An extension of the stability analysis to the so-called *delay-oriented disciplines* can be found in [17].

Further details of the regenerative stability analysis of discrete-time queueing systems, in addition to some refinements, can be found in [4]. More reading on discrete-time queueing models can e.g. be found in [25, 26] and the references therein, while [27, 28] and [29, 30] contain some general reading on queueing models with server interruptions in continuous and discrete time, respectively. It is worth mentioning that the analysis developed in this section can be readily extended to a discrete-time system with a *superposition of independent renewal input processes*. In continuous time, a synchronisation of renewal processes can be achieved if at least one of renewal-time distributions is *spread-out*, meaning that a convolution of this distribution with itself is *absolutely continuous*. However, the main difficulty in the continuous-time setting lies in the construction of a common renewal point of the superposed input process such that the queue size process *simultaneously* achieves a regeneration state, see [5, 31].

A detailed discussion of Jackson networks can be found in [5, 6, 32]. The problem of identifying the stable/unstable stations in some classes of networks can be resolved if equations (10.3.3) are replaced by (see [33, 34])

$$\lambda_i = \lambda P_{0i} + \sum_{j=1}^{N} \min(\mu_j,\, \lambda_j) P_{ji}\,, \quad i = 1, \ldots, N\,.$$

In multiclass networks, different classes of customers, as a rule, follow different input renewal processes, and in this case the *fluid stability analysis* [35] in general turns out to be the most effective. For the interested reader, we mention some additional contributions [15, 36–38], among many, where various aspects of the stability analysis of queueing networks are presented. Also we mention paper [39] in which stability (insensitivity) of the output of a generalized Jackson network is considered.

The applicability of the regenerative method is closely connected to simulation and estimation, and in this regard we mention the books [40, 41] that focus on simulation of regenerative queueing networks, and the papers [42, 43] where the applicability of regenerative simulation is discussed. The theoretical basis of the regenerative simulation method is treated in significant detail in the monograph [44].

There are other general concepts of stochastic processes which inherit important structural properties of classically regenerative processes, for instance, *one-dependent regeneration* (when two adjacent cycles are dependent), and *wide-sense regenerative processes* which maintains the renewal property of the embedded processes and allows a dependence between regeneration cycles. Such a regeneration appears in the stability analysis of recurrent *Harris Markov chains*. An effective approach leading to one-dependent regeneration in queueing systems and networks is based on the theory of *renovating events* developed in [45]. We would like to mention the following important contributions which consider non-classical regeneration [5, 31, 42, 46–56], and furthermore refer to a brief review of various types of regenerations given in the 'Notes' section in [57]. Also we mention related papers [14, 58]. For a *crudely regenerative process*, only the (conditional) expectation of the process is restored at the regeneration instants that constitute a renewal process [59]. In this regard, we already noted in Sect. 2.6 that as long as a renewal process of regenerations can be constructed, the stability analysis of the corresponding model based on the asymptotic characterization of the remaining renewal time can be carried out by the method presented in this book.

References

1. Rogiest, W., Morozov, E., Fiems, D., Laevens, K., Bruneel, H.: Stability of single-wavelength optical buffers. Eur. Trans. Telecomm. **21**, 202–212 (2010)
2. Laevens. K., Bruneel, H.: Analysis of a single-wavelength optical buffer. In: Proceedings 22nd Annual Joint Conference of the IEEE Computer and Communications Societies (INFOCOM 2003), pp. 2262–2267. San Francisco (2003)
3. Rogiest, W., Laevens, K., Fiems, D., Bruneel, H.: Modeling the performance of FDL buffers with wavelength conversion. IEEE Trans. Commun. **57**(12), 3703–3711 (2009)
4. Morozov, E., Fiems, D., Bruneel, H.: Stability analysis of multiserver discrete-time queueing systems with renewal-type server interruptions. Perf. Eval. **68**, 1261–1275 (2011)
5. Asmussen, S.: Applied Probability and Queues, 2nd edn. Springer, New York (2003)
6. Chen, H., Yao, D.D. (edn.): Fundamentals of Queueing Networks: Performance, Asymptotics, and Optimization. Springer, New York (2001)
7. Sherman, N.P., Kharoufeh, J.P.: Optimal Bernoulli routing in an unreliable M/G/1 retrial queue. Prob. Eng. Inf. Sci. **25**, 1–20 (2011)
8. Borovkov, A.A.: Limit theorems for queueing networks I. Theory Probab. its Appl. **31**, 474–490 (1986)
9. Baccelli, F., Foss, S.: Stability of Jackson-type queueing networks. Queueing Syst. **17**, 5–72 (1994)
10. Morozov, E.: Instability of nonhomogeneous queueing networks (2002)
11. Serfozo, R.F.: Basics of Applied Stochastic Processes. Springer, Heidelberg (2009)
12. Morozov, E.V.: Conservation of a regenerative stream in an acyclic net. J. Soviet Math. **57**, 3302–3305 (1991)
13. Morozov, E.: The tightness in the ergodic analysis of regenerative queueing processes. Queueing Syst. **27**, 179–203 (1997)
14. Morozov, E.: Weak regeneration in modeling of queueing processes. Queueing Syst. **46**, 295–315 (2004)

15. Kaspi, H., Mandelbaum, A.: Regenerative closed queueing networks. Stoch. Stoch, Rep. **39**(4), 239–258 (1992)
16. Morozov, E.V.: Regeneration of a closed queueing network. J. Math. Sci. **69**, 1186–1192 (1994)
17. Morozov, E., Rogiest, W., De Turck, K., Fiems, D.: Stability of multi-wavelength optical buffers with delay-oriented scheduling. Trans. Emerg. Telecom. Tech. **23**(3), 217–226 (2012)
18. Chen, Y., Qiao, C., Yu, X.: Optical burst switching: a new area in optical networking research. IEEE Netw. **18**(3), 16–23 (2004)
19. Qiao, C., Yoo, M.: Optical burst switching (OBS)-a new paradigm for an optical internet. J. High-Sp. Netw. **8**, 69–84 (1999)
20. Beheshti, N., Burmeister, E., Ganjali, Y., Bowers, J.E., Blumenthal, D., McKeown, N.: Optical packet buffers for backbone internet routers. IEEE/ACM Trans. Netw. **18**(5), 1599–1609 (2010)
21. El-Bawab, T., Shin, J.-D.: Optical packet switching in core networks: between vision and reality. IEEE Commun. Mag. **40**, 60–65 (2002)
22. Callegati, F., Cerroni, W., Corazza, G.: Optimization of wavelength allocation in WDM optical buffers. Opt. Netw. Mag. **2**(6), 66–72 (2001)
23. Callegati, F., Cerroni, W., Pavani, G.-S.: Key parameters for contention resolution in multi-fiber optical burst/packet switching nodes. In: Proceedings 4th IEEE International Conference on Broadband Communications, Networks and Systems (Broadnets 2007), pp. 217–223. Raleigh (2007)
24. Tancevski, L., Tamil, S., Callegati, F.: Non-degenerate buffers: a paradigm for building large optical memories. IEEE Phot. Tech. Lett. **11**(8), 1072–1074 (1999)
25. Alfa, A.S.: Applied Discrete-Time Queues, 2nd ed.. Springer, New York (2016)
26. Bruneel, H., Kim, B.G.: Discrete-Time Models for Communication Systems Including ATM. Kluwer Academic Press, Boston (1993)
27. Federgruen, A., Green, L.: Queueing systems with service interruptions. Oper. Res. **34**(5), 752–768 (1986)
28. Federgruen, A., Green, L.: Queueing systems with service interruptions II. Nav. Res. Logist. **35**(3), 345–358 (1988)
29. Fiems, D., Steyaert, B., Bruneel, H.: Discrete-time queues with generally distributed service times and renewal-type server interruptions. Perf. Eval. **55**(3–4), 277–298 (2004)
30. Fiems, D., Maertens, T., Bruneel, H.: Queueing systems with different types of interruptions. Eur. J. Oper. Res. **188**, 838–845 (2008)
31. Sigman, K.: One-dependent regenerative processes and queues in continuous time. Math. Oper. Res. **15**, 175–189 (1990)
32. Serfozo, R.F.: Introduction to Stochastic Networks. Springer, New York (1999)
33. Foss, S.: On some properties of open queueing networks. Probl. Inform. Transm. **25**, 90–97 (1989)
34. Goodman, J.B., Massey, W.A.: The non-ergodic Jackson network. J. Appl. Prob. **21**, 860–869 (1984)
35. Dai, J., Hasenbein, J., Kim, B.: Stability of join-the-shortest-queue networks. Queueing Syst. **57**, 129–145 (2007)
36. Afanas'eva, L.G.: On the ergodicity of open queueing network. Th. Prob. Appl. **32**(4), 777–781 (1987)
37. Chang, C.-S., Thomas, J.A., Kiang, S.-H.: On the stability of open networks: a unified approach by stochastic dominance. Queueing Syst. **15**, 239–260 (1994)
38. Meyn, S.P., Down, D.: Stability of generalized Jackson networks. Ann. Appl. Probab. **4**, 124–148 (1994)
39. Morozov, E.: Stability of Jackson-type network output. Queueing Syst. **40**, 383–406 (2002)
40. Shedler, G.S.: Regeneration and Networks of Queues. Springer, New York (1987)
41. Shedler, G.S.: Regenerative Stochastic Simulation. Academic Press Inc., San Diego (1993)
42. Glynn, P.W.: Some topics in regenerative steady-state simulation. Acta Appl. Math. **34**, 225–236 (1994)
43. Glynn, P.W., Iglehart, D.L.: Conditions for the applicability of the regenerative method. Manag. Sci. **39**, 1108–1111 (1993)

44. Asmussen, S., Glynn, P.W.: Stochastic Simulation: Algorithms and Analysis. Springer Nature, Cham (2007)
45. Borovkov, A.A.: Asymptotic Methods in Queueing Theory. Wiley, New York (1984)
46. Asmussen, S., Foss, S.: Renovation, regeneration, and coupling in multiple-server queues in continuous time. Front. Pure Appl. Prob. **1**, 1–6 (1993)
47. Foss, S., Kalashnikov, V.: Regeneration and renovation in queues. Queueing Syst. **8**, 211–224 (1991)
48. Glynn, P.W., Sigman, K.: Uniform Cesaro limit theorems for synchronous processes with applications to queues. Stoch. Proc. Appl. **40**(1), 29–43 (1992)
49. Glynn, P.W.: Wide-sense regeneration for Harris recurrent Markov processes: an open problem. Queueing Syst. **68**, 305 (2011)
50. Kalashnikov, V., Rachev, S.: Mathematical Methods for Construction of Queueing Models. The Wadsworth and Brooks/Cole Mathematical Series, Springer, New York (1990)
51. Nummelin, E.: Regeneration in tandem queues. Adv. Appl. Prob. **13**, 221–230 (1981)
52. Sigman, K.: Regeneration in tandem queues with multiserver stations. J. Appl. Prob. **25**, 391–403 (1988)
53. Sigman, K.: The stability of open queueing networks. Stoch. Proc. Appl. **35**, 11–25 (1990)
54. Sigman, K.: Queues as Harris recurrent Markov chains. Queueing Syst. **3**, 179–198 (1988)
55. Smith, W.L.: Regenerative stochastic processes. Proc. R. Soc. (Ser. A) **232**, 6–31 (1955)
56. Thorisson, H.: Construction of a stationary regenerative process. Stoch. Proc. Appl. **42**, 237–253 (1992)
57. Thorisson, H.: Coupling, Stationarity, and Regeneration. Probability and its Applications, Springer, New York (2000)
58. Morozov, E.V.: Wide sense regenerative processes with applications to multi-channel queues and networks. Acta. Appl. Math. **34**, 189–212 (1994)
59. Serfozo, R.F.: Applications of the key renewal theorem: Crudely regenerative processes. J. Appl. Prob. **29**, 384–395 (1992)

Index

E. Morozov and B. Steyaert, *Stability Analysis of Regenerative Queueing Models*, https://doi.org/10.1007/978-3-030-82438-9

Printed in the United States
by Baker & Taylor Publisher Services